BAMBOO - CURRENT AND FUTURE PROSPECTS

Edited by **Abdul Khalil H.P.S.**

Bamboo - Current and Future Prospects
http://dx.doi.org/10.5772/65539
Edited by Abdul Khalil H.P.S.

Contributors

Mayowa Akeem Azeez, Joshua Iseoluwa Orege, Han Chien Lin, Masazumi Kayama, Takayoshi Koike, Sutarja I Nyoman, Yunhai Ma, Yucheng Liu, Jin Tong, Liu Weiyi, Hui Chaomao, Fang Wang, Dong-Qin Dai, Li-Zhou Tang, Hai-Bo Wang, Guillermo Tardio, Slobodan B. Mickovski, Joao Paulo Fenandes, Johann Peter Rauch, H.P.S. Abdul Khalil, Syifaul Huzni, Sulaiman Thalib, Mustapha Asniza, Chaturbhuj K.S, I.U.H Bhat, Muhammad Ikramullah, Samsul Rizal, Amin Moradbak, Paridah Md Tahir, Folahan, Abdulwahab, Taiwo Owolabi, Ainun Zuriyati Mohamed

First published in London, United Kingdom, 2018 by IntechOpen
IntechOpen is the global imprint of INTECHOPEN LIMITED, registered in England and Wales, registration number: 11086078, The Shard, 25th floor, 32 London Bridge Street
London, SE19SG – United Kingdom
Printed in Croatia

British Library Cataloguing-in-Publication Data
A catalogue record for this book is available from the British Library

Additional hard copies can be obtained from orders@intechopen.com

Bamboo - Current and Future Prospects, Edited by Abdul Khalil H.P.S.
p. cm.
Print ISBN 978-1-78923-230-1
Online ISBN 978-1-78923-231-8

For EU product safety concerns:
IN TECH d.o.o., Prolaz Marije Krucifikse Kozulić 3, 51000 Rijeka, Croatia,
info@intechopen.com or visit our website at intechopen.com.

Meet the editor

Prof. Abdul Khalil, HPS, obtained his PhD degree from the University of Wales, Bangor, United Kingdom, in 1999. He is currently a senior professor in biomass and biocomposites at the School of Industrial Technology, Universiti Sains Malaysia, Penang, Malaysia. His fields of specialization cover biocomposite technology, fibre-reinforced polymer composite technology, hybrid composites, nano-based composites, fibre science (natural and man-made fibre) and biodegradable polymer as new materials for new economy. He is a fellow of Institute of Materials, Minerals and Mining (FIMMM), United Kingdom. He is a member of America Chemical Society and a member of Wood Science and Technology. He had been appointed as an editorial board of more than five Scopus/ISI listed journals. To date, he has generated over 200 scientific articles in Scopus with H-Index of 35 and over 5,000 citations. He also excellently published 3 academic books, 41 chapters in books and also 4 nonacademic books (community/coffee table books). As an expert researcher in his field, he had reviewed more than 200 articles in reputed journals. Recently, he was also awarded as one of the Top Research Scientists Malaysia from Academy of Sciences Malaysia in 2014 and Malaysia's Rising Star Award (Thomson & Reuters) in 2015.

Contents

Preface

Recent development in bamboo technology in the area of construction, furniture, pulp and paper, etc., has proven the positive advantages of bamboo for innovative design and product development in various applications, especially in building construction, furniture, biocomposite products, automotive and environmental management. Optimum utilization of bamboo requires innovative process technology to realize the quality product development. Many value additional attempts have been made on bamboo to enhance the quality through regular maintenance and preservation.

This book was prepared and published not only to draw attention to the potential of bamboo species as it were but to chronicle the various research works carried out on bamboo fibres from different extraction means with the aim of enhancing value addition to the bamboo species. For about four decades of research and development of many bamboo species, the area of research interest has been revolving around the identification, specification, taxonomy and classification of various species and general of bamboos, which are kept in herbaria. It is worth mentioning that bamboo has been used over many decades for housing construction, clothing and other exotic luxuries. The astonishing, ecstatic, beauty resourcefulness and potency of bamboo have attracted wide cultural and sundry applications that have endowed bamboo by giving it a larger role in human culture than any other plant.

This book reports mainly on the various data and information gathered from the recent efforts and milestone achieved in the course of the research. The book sums up the unlimited exploitation of bamboo for industrial purposes as to expose readers on a wide range of potential uses of bamboo nowadays. It is not only offered as an introduction to anyone starting out in research on bamboo-based products but also provides the background for further reading. Chapters in this book discuss, but are not limited to, the bamboo cultivation, properties and applications as well as modification and product developments. The whole book is captivating because it is comprehensive with critical approach to meeting new challenges of technology and innovations.

I am highly thankful to all authors who contributed book chapters and provided their valuable innovative ideas and knowledge. I greatly appreciate contributors' commitment and support towards the completion of this edited book. This book would not have appeared without the help and assists of IntechOpen team. I am very thankful to the team for their generous cooperation at every stage of the book production.

Abdul Khalil H.P.S.
Professor, School of Industrial Technology
Universiti Sains Malaysia
Penang, Malaysia

Section 1

Introduction

Chapter 1

Introductory Chapter: An Overview of Bamboo Research and Scientific Discoveries

Abdul Khalil H.P.S.

Additional information is available at the end of the chapter

http://dx.doi.org/10.5772/intechopen.77094

1. Introduction

The group of plants belonging to a family of giant woody grasses is called bamboo. Among the wide spectrum of uses of bamboo are conceivable purpose, like scaffolding to boats, cooking utensils, furniture, food, fuel, landscaping, ornamental display and a thousand other uses [1]. Despite the enormous importance of bamboo to our social and economic well-being, researchers are still exploiting the potential derivable from most bamboos, in accordance to the International Network for Bamboo and Rattan (INBAR) vision 2015–2030 to carry out sustainable development goals. Furthermore, the food and agricultural organization in collaboration with INBAR (FAO/INBAR) has devised a strategic goal by promoting innovative solutions for poverty and environmental sustainability using bamboo and rattan [2]. However, there is a need to set a direction and define a road map for green development through research and expansion of research breakthroughs using bamboo as a natural fiber source. Being an example of natural fibers, bamboo is renewable, carbon neutral and biodegradable and also produces waste that is either organic or can be used to generate electricity or make ecological housing material [3]. It provides socio-economic benefits, since natural fibers not only come from the environment, but also benefit it.

2. Overview of bamboo research

In order to achieve the above goals, information about the latest empirical breakthroughs with respect to bamboo research scientific discoveries is very pertinent. It is assumed that these findings will further enhance the previous contributions of researchers on the latest scientific applications of bamboo fibers and result of research activities. These are necessary indicators that might be used to assess progress towards these and future milestones. Efforts have been focused in the past towards addressing some of the most pressing contemporary socio-economic issues that bother on the solutions for poverty and environmental sustainability using bamboo. It has been revealed that, apart from the very long species of bamboo, there exist dwarf bamboo which are distributed in areas like northern part of Japan [4]. The growth characteristics of different species of bamboo with respect to geographical locations are therefore important in order to have an idea of the distribution of this dominant forest floor bamboo species in the respective region.

A review of microfungi, i.e., bamboo ascomycetes, with brief appraisal of major morphology and phylogeny of bambusicolous ascomycetes needs to be highlighted, while various species of fungi associated with bamboo need to be enumerated. Moreover, the history, distribution and importance of bamboo ascomycetes to agricultural and economic development as well as research prospects of bamboo ascomycetes will assist in examining the physico-mechanical and thermal studies of bamboo fiber [5]. The knowledge of the fundamental characteristics of bamboo fiber to maximize bamboo utilization in the area of providing mechanical and physical properties of bamboo has been seen as panacea which affects the bamboo durability and strength.

Another area of utmost interest is the modification of bamboo for sustainable product development. Both chemical and thermal modifications of bamboo otherwise called pyrolysis of bamboo enhance the properties of bamboo by extending the bamboo fiber applications [6]. Chemical treatments assist in reinforcing bamboo by modifying the bamboo surface and improve its compatibility with polymer matrix [7]. One of the major by-products of the charcoal collected during the pyrolysis of the charcoal is bamboo vinegars. The potential usage of charcoal from bamboo pyrolysis as safe pigment for food and natural moisture-proof material is another area of paramount research interest. The end use of these modificational approaches on bamboo is for food, energy, pulp, paper, buildings, etc.

The potential of bamboo as a new renewable energy is another breakthrough in the development and scientific cultivation of bamboo. This has been reported to greatly influence people's sociocultural lifestyle and emergence of natural fibre-based composites from various types of bamboo. The influence of type and form of bamboo fibers as fillers and reinforcements has further promoted the performance of bamboo composites. Bamboo production for traditional and modern applications as the country develops and emergence of new economic activities with bamboo resources utilization is very prominent specifically in forest-depleted region like China [8]. Furthermore, the application of bamboo in friction material and as reinforcement materials in concrete structure has been reported as safe and comfortable and could provide simple house for poor families. Recently bamboo is being used as reinforcement in thermoplastic polymers [9], since natural fiber such as bamboo exhibits comparable reinforcement effects to that of conventional glass fibers [10]. The changes in engineering properties of the thermoplastic matrices after the incorporation of untreated and treated bamboo are few of the uses of bamboo in polymer composite. The recent report of the suitability of the alkaline sulphite anthraquinone and methanol pulping on bamboo culms, i.e., *G. scortechinii* bamboo species, has been attracting the attention of polymer scientists [11]. Its use in industrial production of pulp and paper packaging was buttressed by the biometric characterization of the bamboo culms together with the optimum pulping parameters.

3. Conclusion

Since bamboo has high potential socio-economic value, more efforts should be geared towards unraveling possible application of the various forms of bamboo. With the aim of maintaining a good balance between the fundamental and practicability of bamboo utilization, critical approach to meeting new challenges of technology and innovations is a task the researchers should prepare to surmount.

Acknowledgements

The author is thankful to the School of Industrial Technology, Universiti Sains Malaysia (USM), for contributing effectively towards the success of this book.

Author details

Abdul Khalil H.P.S.

Address all correspondence to: akhalilhps@gmail.com

School of Industrial Technology, Universiti Sains Malaysia, Penang, Malaysia

References

[1] Abdul Khalil HPS, Bhat IUH, Jawaid M, Zaidon A, Hermawan D, Hadi YS. Bamboo fibre reinforced biocomposites: A review. Materials and Design. 2012;**42**:353-368

[2] Liese W. The Anatomy of Bamboo Culms. Beijing, China: INBAR Technical Report (China); 1998

[3] Shmulsky R, Jones PD. Forest Products and Wood Science. West Sussex, United Kingdom: John Wiley & Sons; 2011

[4] Parkkeeree T, Matan N, Kyokong B. Flattening of half tubular bamboo culms and fixation of bamboo boards. Journal of Tropical Forest Science. 2014;**26**(1):101-114

[5] Grosser D, Liese W. On the anatomy of Asian bamboos, with special reference to their vascular bundles. Wood Science and Technology. 1971;**5**(4):290-312

[6] Suhaily SS, Abdul Khalil HPS, Nadirah WW, Jawaid M. Bamboo based biocomposites material, design and applications. In: Materials Science-Advanced Topics. Rijeka, Croatia: InTech; 2013. pp. 489-517

[7] Abdul Khalil HPS, Hossain MS, Rosamah E, Norulaini NN, Leh CP, Asniza M, Davoudpour Y, Zaidul ISM. High-pressure enzymatic hydrolysis to reveal physicochemical and thermal properties of bamboo fiber using a supercritical water fermenter. BioResources. 2014;**9**(4):7710-7720

[8] Hao RM, Liu WJ, Chen XY. Investigation on bamboo chip bending process by microwave-heating softening (in Chinese). Journal of Central South University of Technology. 2014;**34**(8):111-113

[9] Huang M, Zhang W, Zhang X, Yu W, Li W, Liu X, Dai C, Wang S. Factors for phyllostachys edulis timber glass transition temperatures (in Chinese). Journal of Zhejiang Agricultural University. 2015;**32**(6):897-902

[10] Tokoro R, Vu DM, Okubo K. How to improve mechanical properties of polylactic acid with bamboo fibers. Journal of Materials Science. 2008;**43**:775-787

[11] Moradbak A, Tahir PM, Mohamed AZ, Halis RB. Alkaline sulfite anthraquinone and methanol pulping of bamboo (*Gigantochloa scortechinii*). BioResources. 2015;**11**(1):235-248

Section 2

Extraction and Modification

Chapter 2

Alkaline Sulfite Anthraquinone and Methanol (ASAM) Pulping Process of Tropical Bamboo (*Gigantochloa scortechinii*)

M.T. Paridah, Amin Moradbak, A.Z. Mohamed,
Folahan Abdulwahab Taiwo Owolabi,
Mustapha Asniza and Abdul Khalil H.P.S.

Additional information is available at the end of the chapter

http://dx.doi.org/10.5772/intechopen.76806

Abstract

This chapter explores the characteristic potentials of alkaline sulfite anthraquinone and methanol (ASAM) pulping of bamboo culms (Gigantochloa scortechinii) in the industrial production of pulp and paper for packaging. The biometric characterization results of the bamboo culms show that bamboo has fiber length of 1980–4000 μm, Runkel ratio of 0.86, and flexibility ratio of 50.19, while the chemical compositions of the bamboo contain 47.67% cellulose, 68.33% holocellulose, 26% lignin, and 3.69% solvent extractive, which give good paper quality fiber and also falls within the range of wood from softwoods species. The study revealed that the optimum ASAM pulping parameters was at 16% NaOH and 90 min cooking time, resulting in Kappa number of 14.17 and pulp yield of 49.06%, while the paper tensile index of 20.86 Nm/g, tear index of 22.64 mN.m2/g, and brightness of 39.32% were obtained. The biometric and chemical characterizations of the ASAM pulped bamboo have shown that ASAM pulped bamboo produces high-quality pulp and paper suitable for packaging and printing paper. Hence, the use of bamboo materials can reduce the burden on the forest, due to the increasing demand for paper and paper products, while supporting the natural biodiversity.

Keywords: alkaline sulfite anthraquinone and methanol, ASAM, pulping, bamboo, pulp and paper

1. Introduction

Despite the global digital revolution and consequence to the variation in the global regional market shares with respect to the production capacity and consumption pattern, the pulp and paper industry has experienced dramatic growth in the recent years. The recent trend of packaging that is gaining consumer's attraction is the emergence of paper and paperboard packaging due to its effective solutions in food and beverages, healthcare, manufacturing, personal care and other industries [1]. The demand of innovative and convenient packaging features is synonymous to the changing lifestyle of people. Conversely, the growth of the pulp and paper industries has been hindered since the major raw material for pulp and paper manufacturing is wood, which is drastically depleting. The industries are saddled with various challenges from ecological protocols, energy consumption and economic survival; therefore, modified technologies that are environmentally benign and economically viable are being sought to enhance the survival of the industries through the twenty-first century and beyond. Various researchers have reported the replacement of wood with non-wood as the immediate panacea to the challenges in the pulp and paper industries. In the array of these non-woods is bamboo, which is a fast-growing monocotyledon plant found in temperate, subtropical and tropical areas with diverse groups of plant in the grass family with 75 genera and 1250 species of bamboo worldwide [2]. Conventionally, bamboo has gained commercial usage in the field of construction (poultry cages, shade blinds,), furniture and handicrafts (vegetable baskets, incense sticks, tooth picks, chopsticks, skewers, barbeque sticks and joss paper) [3].

1.1. Socio-economic benefits of bamboo

Traditionally, being one of the oldest materials used worldwide with wide applicability, bamboo industry is a flourishing business especially at the small-scale level. Since the engine that drives the growth of any economy is the small-scale industries, bamboo business plays a very significant role in the overall growth of an economy. Statistics revealed that of all forests in the world, bamboos occupy approximately 36 million hectares at the rate of present expansion of 3.2% of total forest area [4]. Bamboo can be divided into four regions based on supply chain in various continents of the world and this include Asia-Pacific bamboo region (India, China, Indonesia, Myanmar and Malaysia) [5, 6], which contributes about 65% global supply of bamboo. This is closely followed by American bamboo region (North American and Latin American countries) which contributes about 28% and African bamboo region which contributes about 7% supply. Asia-Pacific bamboo region is expected to continue to dominate the global bamboo market both in terms of value and volume by 2027. Consequently, the cumulative average growth rate (CAGR) of bamboo is projected at 11.8% CAGR. Hence by the end of 2027, nearly three-fourth of global bamboo revenue will be accounted by the region [7]. This projection is based on the versatility of bamboo entrepreneurship, which ramifies wood and furniture (timber substitute, plywood, mat boards, flooring, furniture, outdoor decking), construction (scaffolding, housing, roads), food (bamboo shoots), pulp and paper, textile, agriculture, charcoal and handicrafts, regarded as being one of the key instruments in uplifting the socio-economic status of the poor and underprivileged people in a community. Being the largest consumer of lignocellulose biomass, pulp, paper and paper packaging industries are expected to benefit from this wide availability of bamboo.

2. Suitability of bamboo for pulp and paper

Amid the collection of non-wood natural fibers, bamboo has gained wide applicability in pulp and paper industry. Several reports on the potential of bamboo revealed its potential as a better raw material for pulp and paper packaging because of its long fiber, since the fiber length is an important factor in paper making. The fiber length is similar with what is obtained in softwood fiber properties, hence making bamboo as an ideal replacement in pulp and paper manufacture. For instance, the fiber length has impacts on the physical and mechanical properties of paper such as tear strength and folding endurance. These spectra of properties make bamboo very economical in terms of management and can easily be transported to the pulp and paper mills [8].

The large lumen width and fiber diameter characteristic of bamboo contribute positively to the effective beating process [9]. Like other wood species bamboo fibers have many similar benefits than wood such as: (1) bamboo can be chipped in a similar manner to wood, (2) bamboo is a low-cost crop due to low maintenance, (3) bamboo chips can be blended with wood chips and (4) bamboo is a fast-growing fiber source. Almost all studies suggested that papers made from bamboo are relatively stronger than those made from softwood. Successful development of bamboo pulping requires a technology which is able to exploit the full potential of this raw material. Specifically, this means that selectivity in pulping and bleaching has to be on a high level resulting in pulps with high yield and good strength properties.

2.1. Biometric characteristics of *G. scortechinii*

Table 1 shows the biometric characterization of the studied bamboo (*G. scortechinii*) species compared with some other natural fibers for pulp and paper making. From the table, it means that the bamboo species under study possesses relatively long fibers, 1980 μm, compared to Eucalyptus (840 μm), *B. tulda* (1890 μm) and cotton stalks (810 μm).

Many studies have reported that the fiber length of bamboo is generally greater than hardwood and is similar to softwood [14]. Fiber length may actually be a detriment to good strength properties, as in the case of certain non-wood fibers, such as bamboo, bagasse and cotton. In such a case, fibers are longer than 2 μm [15]. The long fibers are covered with fibrils, fines, and the large fibrils effectively serve to bridge the gaps between the naturally rough surfaces of adjacent fibers. In this way, long fibers have an effect on the strength of a bond and especially its toughness, which is characteristic of the bonds between natural paper making fibers and which is completely lacking with bonds between smooth viscose fibers. On the other hand, water drains from long fiber pulps more rapidly, and this is a point in their favor. Therefore, for pulp and paper production, species with higher fiber lengths are preferred since a better fiber net can be achieved, resulting in a paper with high resistance. The biometric characteristic of *G. scortechinii* shows favorable properties as a fiber raw material for pulp and paper.

Biometric parameters	Type of fibers				
	G. scortechinii	*Bambusa tulda*	*Eucalyptus grandis*	Cotton stalk	Spruce
FL (μm)	1980 ± 3.3	1890	840	810	19
FD (μm)	17.27 ± 3.7	3.45	10.1	16.75	10
LD (μm)	8.66 ± 2.3	6.78	4.4	4.12	4.5
CWT (μm)	0.86 ± 0.20	3.93	0.87	0.49	0.9
Runkel ratio	50.14 ± 3.6	20.29	53.15	67.05	52.63
Flexibility ratio	114.64 ± 3.6	111.2	44.21	32.42	152.63
Slenderness ratio	This study	[10]	[11]	[12]	[13]

FL = fiber length, FD = fiber diameter, LD = lumen diameter, CWT = cell wall thickness, Runkel ratio = (2 × CWT)/LD, Flexibility ratio = (LD/FD) × 100, Slenderness ratio = (FL/FD).

Table 1. Biometric characteristics of bamboo and other natural fibers.

The average fiber diameter of *G. scortechinii* is 17.27 μm which is in the range of other species of Bambusa genera. The lumen diameter and cell wall thickness of bamboo was 8.66 μm and 3.74 μm, respectively. The lumen diameter and cell wall thickness influence the beatability of fibers. Fibers with large lumen diameter and thin walls such as in bamboo have higher bonding abilities due to the better penetration by water into the cell wall and into the lumen causing the cells to swell. The combined effects of the beating action and swelling cause the bonds between the structures of lamellas to loosen and easily separate [16].

2.2. Chemical compositions

The chemical compositions of *G. scortechinii* are shown in **Table 2**. The results showed that the *G. scortechinii* had high contents of ash and silica (1.98 and 1.56%, respectively), but the results were within the range of tropical hardwood, which is 1–3% [20].

The mechanical properties of final paper are related to the amount of hemicellulose and cellulose contents of raw materials. In addition, polysaccharides such as cellulose and hemicellulose have a positive effect on pulp yield of alkaline pulping process [21]. Raw material with more than 34% of cellulose content can be used for pulp and paper production [22]. The cellulose content of *G. scortechinii* is in the range of both soft and hardwood, 40–52% and 38–56%, respectively [23]. The cellulose molecule has numerous hydroxyl groups, which have a strong tendency toward hydrogen bonding with hydroxyl of adjacent molecules [24]. This tendency in connection with the molecular structure is responsible for increasing fiber-to-fiber bonding in final paper [25, 26]. The free hydroxyl groups of cellulose have a strong affinity for polar solvents and solutes that can reach them. An example of this type of interaction is the swelling of cellulose with water. During swelling, the hydrogen bonds between cellulose molecules are broken and replaced by hydrogen bonds between the cellulose molecular and water. Therefore, the cellulose content in the surface of fibers formed with water was a true chemical hydrate and, being glue-like, caused the strength of the resulting paper to increase as well [27].

Composition %	**Natural fibers**			
	G. scortechinii	***P. bambusoides***	**Rice straw**	***P. orientalis***
Holocellulose	68.33 ± 3.7	70.50	70.90	74.46
Cellulose	47. 67 ± 3.4	43.30	48.20	44.31
Lignin	26. 00 ± 2.3	24. 50	17.20	25.20
Solvent extraction	3.64 ± 3.2	3.90	3.50	3.40
1% NaOH	19.82 ± 3.5	25.10	49.20	10.26
HWS	5.53 ± 1.2	6.50	16. 20	2. 81
CWS	4. 61 ± 1.4	—	10.70	1.47
Ash	1.98 ± 0.33	1.40	16.60	0.32
Silica	1.56 ± 0.25	—	14.90	—
Reference	Present study	[17]	[18]	[19]

SE = solvent extractives (ethanol-benzene), HWS = hot water solubility, CWS = cold water solubility, and 1% NaOH = 1% sodium hydroxide solubility.

Table 2. Comparison of the chemical compositions of bamboo with other natural fibers.

The lignin content of *G. scortechinii* is 26%, which is less than the tropical hardwood. Also, *G. scortechinii* contains about 3.68% solvent extractive which is quite similar to softwood (3%) but is less than solvent extractive reported for hardwood. *G. scortechinii* has high 1% NaOH solubility (19.82%). 1% NaOH solubility of *G. scortechinii* shows that more low molecular weight components such as hemicellulose could be solved during the alkaline degradation. As a result, the pulp yield of alkaline pulping of bamboo could be decreased and consumption of alkali charge could be increased due to this property [28, 29]. The hot and cold water solubility of bamboo culms were 5.53 and 4.61%, respectively. Extraction with hot water removes carbohydrate materials such as starches. Extraction with cold water removes sugars, gums, tannins, and inorganic compounds [30]. As it is shown in **Table 2** the ash content of *G. scortechinii* was 1.98%. The ash content of bamboo culms was higher than the aspen and white oak with values of 0.43 and 0.87%, respectively [31]. The silica content of *G. scortechinii* was 1.56%, which is low compared with non-wood species such as rice (14.9%) and wheat straw (4.35%). Silica results in many problems such as blunting of saw tooth, scaling during chemical recovery of black liquor and covering the outer surface of fiber during cooking. On the other hand, the silica is dissolved in alkali solutions; therefore, the consumption of alkali charge in cooking liquor is increased [32].

3. Pulping of bamboo (*G. scortechinii*)

Traditionally, kraft pulping has been the most commonly reported pulping milieu for bamboo species [33]. This is due to its strong bundle sheathes, impenetrable epidermis, a complete absence of ray cells and limited area of conducting tissues. As good as this pulping technique is, it is saddled with a lot of environmental legislative issues. The present paradigm shift in the pulp and paper making toward environmental benign pulping has informed the search for better pulping method for bamboo species.

3.1. Drawbacks of the conventional commercial pulping technique

The conventional commercial pulping procedures of annual plants include both kraft and soda pulping. The major drawback of these pulping protocols is that in the case of alkaline cooking liquors, apart from the delignification process, significant decomposition of carbohydrates portion of the lignocellulose biomass is observed by peeling-off reactions and alkaline hydrolysis. Optimum alkaline cooking condition is pertinent as excess alkaline loading could result in lowering of the ISO brightness due to the concept of alkaline darkening. Furthermore, sodium hydroxide easily dissolves phenolic lignin structures; besides, lignin undergoes condensation reactions under strong alkaline conditions which reduce the reactivity of residual lignin. In the like manner, strong alkaline cooking liquors dissolve high silica content of annual plants and precipitates evaporation problems in soda recovery boilers and in the causticizing plant [34].

In contrast, conventional kraft pulping has been used commercially for paper packaging from annual plants such as wood and some bamboo species, due to good paper properties. Kraft pulping has been found suitable for all ranges of fiber sources. It gives paper of high tensile strength and also high efficiency of the recovery of cooking chemicals [35]. In non-phenolic lignin moieties only β-ethers are cleaved. This has also an impact on bleaching. Among the disadvantages of kraft pulping process, low yield pulp, high consumption of bleaching chemicals, emission of obnoxious odor from the pulping process and large mill size require a tremendous capital investment.

Compared to soda pulps, soda/anthraquinone (soda/AQ) pulps have slightly better bleaching ability, higher yield and sometimes better strength. Similarly, alkali sulfite pulping had been known to give good strength properties and significantly higher yields than kraft process [36]. The addition of anthraquinone helped to reduce Kappa number from the unbleached pulp.

In order to overcome the abovementioned challenges, Patt and Kordsachia [37] discovered that the addition of methanol or ethanol extended delignification to levels below kraft or sulfite pulps. They named their process as alkali-sulfite-anthraquinone methanol (ASAM). Unlike the conventional pulping protocols, ASAM pulping process has dual advantages of paper properties and higher pulp brightness. In addition, ASAM pulping results in pulp with high pulp yield, low Kappa number and high paper strength. Furthermore, the obnoxious odor from methyl mercaptan that is generated during kraft pulping is completely absent in ASAM pulping.

3.2. Mechanism of alkaline sulfite anthraquinone and methanol (ASAM)

Alkaline sulfite anthraquinone with methanol pulping known as the ASAM process was developed by Patt and Kordsachia in 1986 as an alternative cooking process option for soda and kraft pulping processes. According to the authors, the chemical materials that make up the ASAM process are sodium sulfite (Na_2SO_3), sodium hydroxide (NaOH), anthraquinone (AQ) and methanol (CH_3OH). These materials play a unique role in the pulping milieu because, sodium sulfite and sodium hydroxide are the major ingredients of the alkaline sulfite process while the anthraquinone and methanol act as catalysts to enhance the chemical penetration delignification of the lignocellulose biomass. **Table 3** shows the characteristics of cooking parameters in the ASAM pulping process, while **Table 4** gives a summary of the key role of each chemical constituent in ASAM blend.

The mechanism of ASAM pulping process can be viewed with respect to the cooking parameters.

3.2.1. *Effect of sodium hydroxide addition in ASAM pulping*

Sodium sulfite and sodium hydroxide are two major active constituents in ASAM pulping process playing the dual role of delignification and increasing pulp brightness. Several research trials toward the effect of replacement of sodium hydroxide with sodium carbonate

Cooking parameters	Value
Total alkali (NaOH)	20–25 o.d
Alkali ratio (Na_2SO_3/NaOH) or (Na_2CO_3/NaOH)	80/20–70/30
Anthraquinone (AQ)	0.075–0.1%
Methanol	10–15 by vol
Temperature	175–180°C
Cooking time at maximum temperature	120–180 min

o.d = oven dry.

Table 3. Characteristic cooking parameters of ASAM pulping process. Source [38].

Cooking parameters	Significance
Total alkali (NaOH)	Accelerant delignification Increases the cooking PH
Alkali ratio (Na_2SO_3/NaOH) or (Na_2CO_3/NaOH)	Sodium hydroxide accelerant delignification in comparison with sodium carbonate Sodium sulfite changed lignin to solvable material Sodium carbonate generates pulp with the strongest tensile index, highest pulp yield, and highest brightness than sodium hydroxide
Anthraquinone (AQ)	Enhances the delignification rate Act as effective stabilizer in both wood and non-wood polysaccharides
Methanol	Enhances chemical penetration into the woody chips Methanol improves the solubility of the AQ

Table 4. ASAM cooking parameters and their role.

or sodium peroxide in the ASAM pulping process have been reported. The advantage of sodium hydroxide in ASAM pulping is that NaOH tends to accelerant delignification than sodium carbonate; furthermore, by applying NaOH as a source of alkali charge, the initial pH of the white cooking liquor enhances leading to the dissolution of more lignin and carbohydrates (hemicellulose) [38].

Apart from the substitution of sodium hydroxide with sodium carbonate in ASAM pulping, attempt has been made with the use of sodium peroxide. It was reported that sodium carbonate substitution gave better pulp products due to the fact that the sodium carbonate-substituted ASAM mixture gave pulp with the stronger tensile index, higher pulp yield and higher brightness [39] due to greater delignification process. This results from the fact that solution of acid soluble lignin contributed to lignin dissolution. The relatively low pH of spent liquor at high temperature results in the partial cleavage of α-ether bonds in lignin [40, 41]. On the other hand, the use of alkaline sodium sulfite as an additive in the ASAM process produces a low Kappa number, high pulp yield and high brightness. In addition, the production cost is also significantly reduced and the total alkalinity consumption becomes less [38].

3.2.2. *Effect of anthraquinone (AQ) addition*

The addition of anthraquinone (AQ) in white cooking liquor increases the delignification rate due to the rate of decrease in the lignin content of wood or non-wood as the alkali consumption during the alkaline pulping process progresses. This can be divided into three phases: (1) initial reaction, (2) bulk delignification and (3) residual delignification.

In the initial reaction phase, alkali is consumed in deacetylation reaction, in neutralization of wood or non-wood acids and in dissolution of readily soluble wood or non-wood carbohydrate components (hemicellulose, tannins, etc.); hence, very little actual delignification occurs. More lignin is eliminated in the period of bulk delignification phase. This elimination occurs much more quickly than the third stage (residual delignification) [41, 42]. As a result, the proportion of lignin removal in polysaccharides happens slowly. Thus, this is only true when a sufficient amount of alkali charge is available. Thereupon, by adding AQ to the cooking liquid, more alkalinity can be stored for the bulk delignification volume and more lignin can be removed [38, 42]. On the other hand, AQ has the role of an effective stabilizer in both wood and non-wood polysaccharides. AQ is effective at extremely lower dosage levels of 0.05–0.1% on oven-dry wood, giving good results in most cases [43].

3.2.3. *Effect of methanol addition*

The addition of methanol at the initial pulping stage enhances chemical penetration into the woody chips by increasing the solubility of lignin, subsequently, producing a uniform (homogeneous) chemical performance in the cooking time and also suppresses the dissociation of the inorganic pulping chemicals. The trapped air from the woody cells could be removed by raising the pressure of the digester, thus accelerating the penetration of chemicals into the wood cells [44, 45].

In ASAM pulping, it is important to ensure that alkali has penetrated into the center of the chip prior to rising the temperature above 140°C, since undesirable lignin condensation reactions can occur at high temperatures in the absence of alkali [46, 47]. The use of methanol is preferred over ethanol in ASAM pulping protocol because of the lower boiling point that results in lower energy consumption for recovery. In addition, compared to ethanol, methanol has less viscosity, polarity and surface tension. The low surface tension and increased pressure, together with the ability of methanol to dissolve resins, have a positive impact on the penetration of chemicals into the chips [48, 49].

Apart from enhancing chemical penetration during pulping, Sridach [50] discovered that methanol also improves the cellulose stability in the ASAM pulping, hence producing pulp with higher viscosity. This was achieved by suppressing, stopping the reaction or slowing down the transformation of the cellulose aldehydic functional groups into keto groups that initiate the peeling reaction. Due to the use of methanol in the cooking liquor, more penetration of chemicals into the chips occurs. Raising pressure of the digester and more air removal of chips are the two main ways of improving the penetration of chemical materials into wood or non-wood chips. Therefore, adding methanol while an ASAM cooking liquor is in the process of penetrating the chips leads to a higher pressure of the digester. As a result, the entrapped air of the chips is displaced, and chemical materials can penetrate into both wood and non-wood chips. Consequently, more delignification happens and fibers can be easily separated [51].

3.2.4. *Advantages of ASAM pulping*

In comparison with kraft pulping process, the ASAM pulping process has a higher digester pressure of 1.3–1.4 MPa and the cooking temperature of 5–10° [42]. Several studies have considered the application of ASAM in the cooking process of both hard and soft wood. Moradbak et al. [42] reported that E*ucalyptus globulus* had been cooked by ASAM pulping process, which results in 56.9 and 53.6% pulp yield and 14 and 10 Kappa number, respectively. Pulp yield and Kappa number of Pinus sylvestris were found to be from 52.9 to 52.5% and from 31 to 27, respectively [52].

Several reports have shown some of the unique advantages of ASAM pulping process. These include (i) prevention of air pollution, (ii) high delignification rate, (iii) high brightness of pulp, (iv) high pulp yield and (v) ease of pulp bleaching [53].

3.2.5. *Reactivity of ASAM pulping*

A review of the reactivity of alkaline sulfite anthraquinone and methanol (ASAM) pulping process shows that the information available is relatively scarce, fragmented and yet to gain adequate commercial applications. Previous studies have focused on either sodium-anthraquinone (Soda/AQ) pulping or alkaline sulfite-anthraquinone (AS/AQ) pulping. In pulping process, the parameters that are of most significance with respect to delignification and polysaccharide removal are alkali charge and cooking time [42, 47]. In addition, ASAM pulping process has better selectivity in comparison to kraft or soda pulping processes in terms of delignification, which leads to low Kappa number and high viscosity. Few studies have

Raw material	Pulping parameters	Total yield	Kappa no.	Tensile index (Nm/g)	Burst index ($kPa.m^2/g$)	Tear index ($mN.m^2/g$)	Brightness (%)	Reference
Bamboo (*G. scortechinii*)	NaOH: 16%; AQ: 0.1% Methanol: 15% Na_2SO_3/NaOH: 80/20 T: 90 min; Temp: 170°C	49.06	14.17	20.86	10.83–9.27	20.86	39.32	Present Study
Hibiscus cannabinus	NaOH: 17%; AQ: 0.1% Na_2SO_3/NaOH: 70/30 T: 120 min; Temp: 175°C	52.6	15.5	101.9	7.4	9.4	41.9	[54]
Whole jute plant	NaOH: 22%; AQ: 0.1% Na_2SO_3/NaOH: 80/20 Methanol: 25% T: 60 min; Temp: 170°C	56.6	39.2	—	3.2	11.2	45	[55]
Eucalyptus globulus	NaOH: 22.5%; AQ: 0.1% Methanol: 25% Na_2SO_3/NaOH: 80/20 T: 135 min; Temp: 180°C	56.7	16.6	53.8	—	4.2	56.6	[56]
Populus deltoids clone	0.1%; Methanol: 10% Na_2SO_3/NaOH: 80/20 T: 120 min; Temp: 170°C	51	—	59.4	4.8	3.8	45.9	[57]
Eucalyptus globulus labill	NaOH: 25%; AQ: 0.1% Methanol: 30% Na_2SO_3/NaOH: 70/30 T: 90 min; Temp: 170°C	53.4	17	—	—	—	36	[58]
Trema orientalis	NaOH: 17%; AQ: 0.1% Methanol: 20% Na_2SO_3/NaOH: 80/20 T: 120 min; Temp: 180°C	52.8	13.4	47.4	4.8	10.7	—	[59]

AQ = anthraquinone, T = time, Temp = temperature.

Table 5. Effects of ASAM pulping parameters on pulp and paper properties of various types of lignocellulosic materials.

considered the application of ASAM in the cooking process of both hard and soft wood. **Table 5** gives the summary of the reported pulping of lignocellulose biomass using ASAM pulping process on different lignocellulose biomass compared with the ASAM pulped bamboo (*G. scortechinii*) species. At a constant ASAM pulping process at 170° and alkali ratio (Na_2SO_3/NaOH; anthraquinone: methanol) of (80/20, 0.1%: 15%), it was found that ASAM pulped bamboo gave a characteristic high yield of 52.36%. Like the other ASAM pulped biomass (**Table 5**) this is quite higher than what is obtainable in kraft pulping. The highest yield was achieved by applying 14% sodium hydroxide and 90 min cooking time while the lowest Kappa number (10.38) was observed using 18% and 120 min. Further increase of both alkali and cooking time resulted in marked decrease in both Kappa number and pulp yield. From the result, ASAM pulping of bamboo (*G. scortechinii*) revealed a low Kappa number compared with what was obtained from *Hibiscus cannabinus*, whole jute plant, *Eucalyptus globulus* and *Eucalyptus globulus labill* as reported by Khristova et al. [54]; Jahan et al. [55]; Kordsachia et al. [56]; and Gominho et al. [58], respectively, despite the high ASAM pulping parameters used.

The tear, tensile and burst indices of unbleached bamboo ASAM paper were within the range of 26.33–18.64 $mN.m^2/g$, 24.8–17.87 Nm/g and 10.83–9.27 $kPa.m^2/g$, respectively, which is suitable for packaging paper. The study revealed that the optimum ASAM pulping parameters were at 16% NaOH and 90 min cooking time, resulting in paper having tensile index of 20.86 Nm/g, tear index of 22.64 $mN.m^2/g$ and brightness of 39.32%. Another advantage of this pulping process is to produce pulp with a low proportion of rejected material. Aldonic acid, which is present as a result of isolation from pulp hydrolyzates, indicates that stabilization takes place through conversion of the end group to the acid out of an oxidation reaction [42, 60]. Therefore, according to role of AQ and methanol in ASAM pulping process, one can expect high yields of ASAM bamboo unbleached pulp.

4. Conclusion

The study determined the technical feasibility of bamboo culms (*G. scortechinii*) as a non-wood fiber for pulp and paper industry. For this purpose, the biometric characteristics and chemical compositions, pulping characteristics, paper properties, projects ASAM pulped of *G. scortechinii* as good material for pulp and paper packaging material. The results show that the fiber length of *G. scortechinii* is similar to softwood and had significant effects on the properties of bamboo paper. The Runkel and flexibility ratios of *G. scortechinii* were in the range of spruce (softwood) with 0.86 and 50.14, respectively. Meanwhile, the chemical composition analysis of *G. scortechinii* presented that bamboo have a high amount of cellulose content and lowest solvent extractive content in comparison with other non-wood species. Therefore, the result showed that *G. scortechinii* can produce pulp and paper, which is almost comparable to other sources. It was also found that the pulp properties of bamboo often being subjected to ASAM pulping process were significantly affected (at $p \le 0.05$) by the cooking conditions. The pulp at 18% NaOH and 90 min cooking time gave low pulp yield and the tear index with values of 41.24% and 18.64 $mN.m^2/g$, respectively. The pulp made at 16% NaOH and 90 min cooking time presented the best properties, pulp yield, Kappa number, tensile, tear and burst indices with values of 49.06%, 14.2, 20.86 Nm/g, 22 $mN.m^2/g$ and 10.05 $kPa.m^2/g$, respectively.

Acknowledgements

The authors are thankful to the Institute of Tropical Forestry and Forest Products (INTROP), Universiti Putra Malaysia (UPM), and the Higher Institution Centre of Excellence (HICoE) Ministry of Higher Education Malaysia, Grant No. 6369107. Additional thanks are extended to the School of Industrial Technology, Universiti Sains Malaysia (USM) and the Federal Institute of Industrial Research Oshoidi Nigeria for technical support.

Author details

M.T. Paridah[1]*, Amin Moradbak[1], A.Z. Mohamed[1], Folahan Abdulwahab Taiwo Owolabi[2,3], Mustapha Asniza[2] and Abdul Khalil H.P.S.[2]

*Address all correspondence to: parida.introp@gmail.com

1 Institute of Tropical Forestry and Forest Products (INTROP), Universiti Putra Malaysia, Serdang, Selangor, Malaysia

2 Bioprocessing and Paper Coating Division, Penang, Malaysia

3 Pulp and Paper Technology Division, Federal Institute of Industrial Research Oshodi, Ikeja, Lagos, Nigeria

References

[1] Bras J, Saini S. Nanocellulose in functional packaging. In: Cellulose-Reinforced Nanofibre Composites. Woodhead Publishing Series in Composites Science and Engineering; 2017. pp. 175-213

[2] Chen J, Yuan Z, Zanuso E, Trajano HL. Response of biomass species to hydrothermal pretreatment. In: Hydrothermal Processing in Biorefineries. Cham: Springer; 2017. pp. 95-140

[3] Chakraborty S, Chaudhuri PS. Earthworm communities in the bamboo plantations of West Tripura (India). Proceedings of the Zoological Society. Springer India. 2017;**70**(2):105-118

[4] Wang Y, Bai S, Binkley D, Zhou G, Fang F. The independence of clonal shoot's growth from light availability supports moso bamboo invasion of closed-canopy forest. Forest Ecology and Management. 2016;**368**:105-110

[5] Banik RL. Bamboo silviculture. In: Bamboo. Springer International Publishing; 2015. pp. 113-174

[6] Ling M, Christensen M, Donnison A, Belmonte KD, Brown C. Scoping Study to Inform the Global Assessment of Bamboo and Rattan (GABAR). Cambridge, UK: UN Environment World Conservation Monitoring Centre. 2016

[7] Catacutan DC, van Noordwijk M, Nguyen TH, Öborn I, Mercado AR. Agroforestry: Contribution to Food Security and Climate-Change Adaptation and Mitigation in Southeast Asia. White Paper. Bogor, Indonesia: World Agroforestry Centre (ICRAF) Southeast Asia Regional Program; Jakarta, Indonesia: ASEAN-Swiss Partnership on Social Forestry and Climate Change. 2017

[8] Dai L, Long Z, Chen J, An X, Cheng D, Khan A, Ni Y. Robust guar gum/cellulose nanofibrils multilayer films with good barrier properties. ACS Applied Materials & Interfaces. 2017;**9**(6):5477-5485

[9] Akeem Azeez M, Andrew JE, Sithole BB. A preliminary investigation of nigerian Gmelinaarborea and *Bambusa vulgaris* for pulp and paper production. Maderas. Ciencia y Tecnología. 2016;**18**(1):65-78

[10] Sharma AK, Dutt D, Upadhyaya JS, Roy TK. Anatomical, morphological, and chemical characterization of *Bambusa tulda, Dendrocalamus hamiltonii, Bambusa balcooa, Malocana baccifera, Bambusa arundinacea* and *Eucalyptus tereticornis*. BioResources. 2011;**6**(4):5062-5073

[11] Malik RS, Dutt D, Tyagi CH, Jindal AK, Lakharia LK. Morphological, anatomical and chemical characteristics of Leucaena leucocephala and its impact on pulp and paper making properties. Journal of Scientific and Industrial Research. 2004;**63**:125-133

[12] Tutus A, Ezici AC, Ates S. Chemical, morphological and anatomicalproperties and evaluation of cotton stalks (*Gossypium hirsutum* l) in pulp industry. Scientific Research and Essays. 2010;**5**(12):1553-1560

[13] He Z. Determination of Strength Properties of Norway Spruce after Sulphate Cooking and Oxygen Delignification. Saimaa University; 2011 Retrieved from http://publications.theseus.fi/handle/10024/30125

[14] Laftah WA, Wan Abdul Rahman WA. Pulping process and the potential of using non-wood pineapple leaves fiber for pulp and paper production: A review. Journal of Natural Fibers. 2016;**13**(1):85-102

[15] Yang H, Chen D, van de Ven TG. Preparation and characterization of sterically stabilized nanocrystalline cellulose obtained by periodate oxidation of cellulose fibers. Cellulose. 2015;**22**(3):1743-1752

[16] Gharehkhani S, Sadeghinezhad E, Kazi SN, Yarmand H, Badarudin A, Safaei MR, Zubir MNM. Basic effects of pulp refining on fiber properties—A review. Carbohydrate Polymers. 2015;**115**:785-803

[17] Deniz I, Ates S. Determination of optimum kraft pulping conditions using bamboo (*Pyllotachys bambusoides*). In: 2nd National Black Sea Forestry Congress Proceedings. Artvin, Turkey; 2002

[18] Tutus A, Eroglu H. An alternative solution to the silica problem in wheat straw pulping. Appita Journal. 2004;**57**(3):214-217

[19] Tutus A, Ates S, Deniz I. Pulp and paper production from spruce wood with Kraft and modified Kraft methods. African Journal of Biotechnology. 2010;**9**(11):1648-1654

[20] Marklein AR, Winbourne JB, Enders SK, Gonzalez DJ, Huysen TL, Izquierdo JE, Norton RA. Mineralization ratios of nitrogen and phosphorus from decomposing litter in temperate versus tropical forests. Global Ecology and Biogeography. 2016;**25**(3):335-346

[21] Stoklosa RJ, Hodge DB. Fractionation and improved enzymatic deconstruction of hardwoods with alkaline delignification. Bioenergy Research. 2015;**8**(3):1224-1234

[22] Vargas González F, González Z, Rojas O, Garrote G, Rodríguez Pascual A. Barley straw (Hordeumvulgare) as a supplementary raw material for *Eucalyptus camaldulensis* and pinussylvestris kraft pulp in the paper industry. 2015

[23] Liese W, Tang TKH. Properties of the bamboo culm. In: Bamboo. Springer International Publishing; 2015. pp. 227-256

[24] Li Y, Liu X, Zhang S, Yao Y, Yao X, Xu J, Lu X. Dissolving process of a cellulose bunch in ionic liquids: A molecular dynamics study. Physical Chemistry Chemical Physics. 2015; **17**(27):17894-17905

[25] Belle J, Odermatt J. Initial wet web strength of paper. Cellulose. 2016;**23**(4):2249-2272

[26] Olejnik K, Stanislawska A, Wysocka-Robak A. Properties of natural cellulose fibers and methods of their modification for the purpose of paper quality improvement. In: Handbook of Sustainable Polymers: Structure and Chemistry. Wolczanska, Lodz, Poland: Institute of Papermaking and Printing, Lodz University of Technology; 2016. p. 19

[27] Yazawa K, Ishida K, Masunaga H, Hikima T, Numata K. Influence of water content on the β-sheet formation, thermal stability, water removal, and mechanical properties of silk materials. Biomacromolecules. 2016;**17**(3):1057-1066

[28] Yuan Z, Kapu NS, Beatson R, Chang XF, Martinez DM. Effect of alkaline pre-extraction of hemicelluloses and silica on Kraft pulping of bamboo (NeosinocalamusaffinisKeng). Industrial Crops and Products. 2016;**91**:66-75

[29] Yuan Z, Wen Y, Kapu NS, Beatson R, Martinez DM. A biorefinery scheme to fractionate bamboo into high-grade dissolving pulp and ethanol. Biotechnology for Biofuels. 2017;**10**(1):38

[30] Taflick T, Schwendler LA, Rosa SM, Bica CI, Nachtigall SM. Cellulose nanocrystals from acacia bark–influence of solvent extraction. International Journal of Biological Macromolecules. 2017;**101**:553-561

[31] Mishra S, Mohanty AK, Drzal LT, Misra M, Hinrichsen G. A review on pineapple leaf fibers, sisal fibers and their biocomposites. Macromolecular Materials and Engineering. 2014;**289**:955-974

[32] Yuan Z, Kapu NS, Beatson R, Chang XF, Martinez DM. Effect of alkaline pre-extraction of hemicelluloses and silica on kraft pulping of bamboo (NeosinocalamusaffinisKeng). Industrial Crops and Products. 2016;**91**:66-75

[33] Bilal M, Iqbal HM, Hu H, Wang W, Zhang X. Metabolic engineering and enzyme-mediated processing: A biotechnological venture towards biofuel production–a review. Renewable and Sustainable Energy Reviews. 2018;**82**:436-447

[34] Bajpai P. Pulp and Paper Industry: Chemicals. Elsevier; 2015

[35] Laftah WA, Wan Abdul Rahman WA. Pulping process and the potential of using non-wood pineapple leaves fiber for pulp and paper production: A review. Journal of Natural Fibers. 2016;**13**(1):85-102

[36] Upton BM, Kasko AM. Strategies for the conversion of lignin to high-value polymeric materials: Review and perspective. Chemical Reviews. 2015;**116**(4):2275-2306

[37] Patt R, Kordsachia O. Herstellung von Zellstoffen unter Verwendung von alkalischen Sulfitlosungen mit Zusatz von Anthrachinon und Methanol. Papier. 1986;**40**(10A):V1-V8

[38] Young RA, Akhtar M. Environmentally Friendly Technologies for the Pulp and Paper Industry. John Wiley & Sons, Inc; 1998

[39] Kautto J. Evaluation of two Pulping-Based Biorefinery Concepts. Acta Universitatis Lappeenrantaensis; 2017

[40] Alekhina M, Ershova O, Ebert A, Heikkinen S, Sixta H. Softwood Kraft lignin for value-added applications: Fractionation and structural characterization. Industrial Crops and Products. 2015;**66**:220-228

[41] Yang X, Li N, Lin X, Pan X, Zhou Y. Selective cleavage of the aryl ether bonds in lignin for depolymerization by acidic lithium bromide molten salt hydrate under mild conditions. Journal of Agricultural and Food Chemistry. 2016;**64**(44):8379-8387

[42] Moradbak A, Tahir PM, Mohamed AZ, Peng LC, Halis R. Effects of alkaline sulfiteanthraquinone and methanol pulping conditions on the mechanical and optical paper properties of bamboo (Gigantochloa scortechinii). BioResources. 2016;**11**(3):5994-6005

[43] Kaur D, Bhardwaj NK, Lohchab RK. A study on pulping of rice straw and impact of incorporation of chlorine dioxide during bleaching on pulp properties and effluents characteristics. Journal of Cleaner Production. 2018;**170**:174-182

[44] Singh R, Krishna BB, Kumar J, Bhaskar T. Opportunities for utilization of non-conventional energy sources for biomass pretreatment. Bioresource Technology. 2016;**199**:398-407

[45] Bhutto AW, Qureshi K, Harijan K, Abro R, Abbas T, Bazmi AA, Yu G. Insight into progress in pre-treatment of lignocellulosic biomass. Energy. 2017;**122**:724-745

[46] Wang S, Ru B, Lin H, Sun W, Luo Z. Pyrolysis behaviors of four lignin polymers isolated from the same pine wood. Bioresource Technology. 2015;**182**:120-127

[47] Owolabi AWT, Arniza G, wan Daud W, Alkharkhi AF. Effect of alkaline peroxide pretreatment on microfibrillated cellulose from oil palm fronds rachis amenable for pulp and paper and bio-composite production. BioResources. 2016;**11**(2):3013-3026

[48] Zhao X, Cheng K, Liu D. Organosolvpretreatment of lignocellulosic biomass for enzymatic hydrolysis. Applied Microbiology and Biotechnology. 2009;**82**(5):815

[49] Lehto JT, Alén RJ. Chemical pretreatments of wood chips prior to alkaline pulping-a review of pretreatment alternatives, chemical aspects of the resulting liquors, and pulping outcomes. BioResources. 2015;**10**(4):8604-8656

[50] Sridach W. The environmentally benign pulping process of non-wood fibers. Suranaree Journal of Science and Technology. 2010;**17**(2)

[51] Korotkova E, Pranovich A, Wärnå J, Salmi T, Murzin DY, Willför S. Lignin isolation from spruce wood with low concentration aqueous alkali at high temperature and pressure: Influence of hot-water pre-extraction. Green Chemistry. 2015;**17**(11):5058-5068

[52] Patt R, Kordsachia O. Anforderungen an neue Zellstoffherstellungsverfahren. Wochenblatt für Papierfabrikation. 1988;**116**(7):709-713

[53] Razalli RL, Abdi MM, Tahir PM, Moradbak A, Sulaiman Y, Heng LY. Polyaniline-modified nanocellulose prepared from Semantan bamboo by chemical polymerization: Preparation and characterization. RSC Advances. 2017;**7**(41):25191-25198

[54] Khristova P, Kordsachia O, Patt R, Khider T, Karrar I. Alkaline pulping with additives of kenaf from Sudan. Industrial Crops and Products. 2002;**15**(3):229-235

[55] Jahan MS, Chowdhury DAN, Islam MK. ASAM Pulping of Jute. Pulp & Paper Science and Technology. Alternative Papermaking Fibres; 2003. pp. 252-258

[56] Kordsachia O, Wandinger B, Patt R. Some investigations on ASAM pulping and chlorine free bleaching of eucalyptus from Spain. HolzAlsRoh-Und Werkstoff. 1992;**50**(3):85-91

[57] Mertoglu-Elmas G, Gunaydin K, Ozden O. Environmental friendly alkaline sulfiteanthraquinone-methonal (ASAM) pulping with Rumexcrispus plant extract of woody materials. Journal of Environmental Biology. 2012;**33**(5):941-944

[58] Gominho J, Lopes C, Lourenço A, Simões R, Pereira H. Eucalyptus globulus stumpwood as a raw material for pulping. BioResources. 2014;**9**(3):4038-4049

[59] Jahan MS, Rubaiyat A, Sabina R. Evaluation of cooking processes for Tremaorientalis pulping. Journal of Scientific and Industrial Research. 2007;**66**(10):853-859

[60] Derrien E, Marion P, Pinel C, Besson M. Influence of residues contained in softwood hemicellulose hydrolysates on the catalytic oxidation of glucose to glucarate in alkaline aqueous solution. Organic Process Research and Development. 2016;**20**(7):1265-1275

Chapter 3

Bamboo, Its Chemical Modification and Products

Mayowa Akeem Azeez and Joshua Iseoluwa Orege

Additional information is available at the end of the chapter

http://dx.doi.org/10.5772/intechopen.76359

Abstract

Bamboo, a perennial woody grass belonging to Gramineae family and Bambuseae subfamily, is ubiquitous in many parts of the world. This biomass possesses high potential as a substitute for many lignocellulosic and non-lignocellulosic materials in various capacities of applications owing to its chemical composition as well as its physical properties. Its abundance, chemical composition and numerous applications are reviewed in this work. This chapter also examined some investigated chemical modifications through alkali hydrolysis, acid hydrolysis, coupling to enhance properties of bamboo fibre for specialised applications.

Keywords: bamboo, biomass, fibre, composition, modification, utilisation, resources

1. Introduction

The demand for a substitute to resources associated with environmental problems has brought about a strong interest in the use of raw materials and products that are renewable, sustainable and biocompatible. However, a great deal of research and technical work has been devoted to the use of natural plant fibre. Natural plant fibres have unequivocally contributed to economic prosperity and sustainability in our daily lives [1]. Bamboos, in particular, have attracted special attention owing to their awesome properties including and not limited to sustainability, renewability and biodegradability. They present versatile structure produced by physical and mechanical properties and low specific weight [2]. Properties such as appearance, strength and hardness combined with its rapid growth rate and capacity for sustainable harvesting have made bamboo an attractive substitute in different industrial sectors and these have successively created great opportunities for its development [3]. It has been reported that the density of bamboo varies from 500 to 800 kg/m^3 depending on anatomical structures such as quantity and distribution of fibres around vascular bundles [4], with its maximum density usually obtained from 3 years old culms [5, 6].

Bamboo is a perennial woody grass, which belongs to the family Gramineae and subfamily Bambuseae [7]. It is an evergreen, monocotyledonous (i.e. non-woody) plant, which produces primary shoot without any later secondary growth [8]. It is widely distributed in the world with China as the most extensive bamboo-producing country [9], having a global export volume of 57.3% in 2009 [10] as shown in **Table 1**. Bamboo is found in abundance in Asia and South America [11]. With over 1200 species and 70 genera, bamboo are found in natural forests, semi-exploited stands and vast plantations globally in an area of more than 14 million ha. [8, 12, 13]. Species of bamboo are found in all continents except in Europe [2]. The percentage of world bamboo resources by continent is shown in **Table 2** [14] and the countries with the most abundant bamboo resources are shown in **Table 3** [15]. The area covered by bamboo in Africa has been estimated at 1.5 million ha with about 43 species [10, 16]; that of Myanmar has been estimated at 2.2 million ha and that of India estimated between 3 and 20 million ha [17]. A total of 40 of the 43 species found in Africa are majorly distributed in Madagascar while the remaining 3 species are found in mainland Africa. Ethiopia has over 1 million ha of highland and lowland bamboo resources, which is about 86% of the African bamboo resource that serves as a subsistence material for rural communities [18, 19].

Bamboo is a lignocellulosic biomass from which some value-added products can be obtained. Although data on worldwide production of bamboo products are incredibly unreliable, they do not appear in significant commodity databases [8]. Its use varies from domestic household products to industrial applications, from medicine to nutrition and from toys to aircraft production.

It has 1500 recorded uses and a combined value of internal and commercial consumption of around US $10 billion in the world market [20]. In recent years, the development and utilisation

Country	Volume of global exports (%)
China	57.3
Indonesia	14.8
Vietnam	4.6
EU-27	3.0
USA	1.7
Philippines	1.6
Thailand	1.0
Singapore	1.0
Myanmar	0.8
Malaysia	0.8

Table 1. Top 10 exporter of bamboo globally in 2009 [10].

http://dx.doi.org/10.5772/intechopen.76359

Continent	Percentage of bamboo produced
Asia	65
South America	28
Africa	7

Table 2. Percentage of world bamboo resources by continent [14].

of bamboo have attracted particular attention, not only in paper, textile and food industry but also for construction and reinforcing fibres. In addition, with adequate technology, the stems can be used in the production of cellulose, bio-ethanol and starch [21]. Other bamboo products include bio-methane, flavonoids and functional xylo-oligosaccharides. Some studies have also shown that bamboo resource could be considered as a candidate feedstock of biomass energy for its high growth efficiency [2, 9, 22]. As a cheap lignocellulosic feedstock, bamboo has been adopted for bioenergy production. This adoption is due to its environmental benefits, fast growth and high annual biomass yield. The various uses of bamboo for humans are quite remarkable. Some minor applications include the use of its leaves for medical purposes, fresh edible shoots and culms for timber or as a raw material for pulping [8].

Bamboo is considered an alternative to wood owing to their excellent qualities in physical and mechanical attributes [23]. It attains maturity in 3 years as compared to wood, which takes almost more than 20 years. After maturity, the tensile strength of bamboo is comparable to that of mild steel [15]. It is one of the fastest growing plants [8]. It can grow in areas that are currently non-productive (e.g. on an eroded slope), and its root structure remains intact after harvest, thus, generating new shoots [24]. Bamboo grows in plains, hilly and high altitude mountainous regions, and in most kinds of soils, except alkaline soils, desert and marsh [25].

According to their morphology, bamboos are broadly divided into monopodial (or running) bamboos with 'leptomorph' rhizome system and sympodial (clumping) bamboos with 'pachymorph' rhizome system. These differences in rhizome systems can be because of their adaptations to climate conditions to which the bamboos belong. As a typical forest plant in the tropical and subtropical area, bamboo forest plays significant roles in its biological characteristics and

Country	Bamboo resources in percentage
India	30
China	14
Indonesia	5
Ecuador	4
Myanmar	2
Viet Nam	2
Others	43

Table 3. Countries with the largest bamboo resources in the world [15].

growth habits. Apart from its socio-economical utilisation, bamboo has many environmental benefits [7]. It has some ecological functions on soil erosion control, water conservation, land rehabilitation, carbon sequestration, etc. In China, bamboo forests are recognised as a massive carbon sink in the global cycles of carbon. They have high potentials in carbon fixation, and this is due to the prediction that the carbon stocks in bamboo stands based on previous data for 2050 may get to 1017.64 Tg C [26] or reach 1138.8 Tg C [27].

Some studies have revealed bamboo to produce higher biomass yield than other lignocelluloses crops with a growth rate ranging from 30 to 60 cm/day and height of about 36 m in growing season [28]. The aboveground biomass of bamboo in the Philippines was first reported as 146.8 Mg ha^{-1} $year^{-1}$ (Suzuki and Jacalne [119]). Nath et al. [29] reported that the total aboveground standing biomass of bamboo in northeast India was 42.98 Mg ha^{-1} $year^{-1}$. Hong et al. [30], when comparing the annual biomass yield between bamboo and *Miscanthus* species, reported that of bamboo to range from 5.9 to 49.5 Mg ha^{-1} $year^{-1}$.

2. Chemical composition of bamboo

Cellulose, hemicelluloses and lignin are the three major chemical compositions of bamboo, and they are closely associated in a complex structure [31]. They contribute about 90% of the total bamboo mass. The minor components are pigments, tannins, protein, fat, pectin and ash. Others include resins, waxes and inorganic salts. These constituents play an important role in physiological activity of bamboo, and they are found in cell cavity or special organelles [15]. The chemical composition of bamboo is known to be similar to that of wood, but bamboo has a higher content of minor components compared with wood [32].

Li et al. [33] in their studies reported the chemical composition of bamboo fibre as shown in **Table 4**. Usually, there is variation in the chemical composition of bamboo depending on their age. Notably, cellulose content decreases with increase in age of bamboo. Different authors have investigated different species and bamboo parts. They include bamboo (Kumamoto, Japan) with cellulose content of 47%, hemicelluloses 23% and lignin 28% [9]; bamboo with cellulose 43%, hemicelluloses 15% and lignin 26% [34]; bamboo (Dendrocalamus sp.) with cellulose 47%, hemicelluloses 16% and lignin 18% [35]; bamboo with cellulose 44%, hemicelluloses 30% and lignin 26% [36]; Moso bamboo (*Phyllostachys pubescens* Mazel), with cellulose 46%,

Chemical constituents	Percentage
Cellulose	73.83
Hemicellulose	12.49
Lignin	10.15
Aqueous Extract	3.16
Pectin	0.37

Table 4. Chemical composition of bamboo fibre [33].

hemicelluloses 23% and lignin 26% [37]; bamboo (*Dendrocalamus asper*) with cellulose 41%, hemicelluloses 27% and lignin 27% [38]; Moso bamboo (*Phyllostachys heterocycla*), with cellulose 42–47%, hemicelluloses 22–23% and lignin 23–31% [39]; bamboo, with cellulose 38.4%, hemicelluloses 20.5% and lignin 20.8% [40]; Moso bamboo (*Phyllostachys heterocycla*), with cellulose 37%, hemicelluloses 22% and lignin 24% [41]; bamboo shoots shell fibre (BSSF), with cellulose 23%, hemicelluloses 14% and lignin 11% and bamboo stem and leaf (BSL), with cellulose 21%, hemicelluloses 12% and lignin 12% [42].

3. Chemical modification of bamboo

Many research and technical works have been carried out on the chemical modification of bamboo fibres to improve their properties for specialised applications [43–46]. Chemical modification methods include alkali hydrolysis, acid hydrolysis, coupling, etc.

Alkali hydrolysis is a conventional technique. It is a chemical processing raw cellulose fibre to delignify and to remove the amorphous regions. It creates a rough fibre surface, activates hydroxyl groups and improves the fibre tensile strength. This process involves the initial use of an alkali solution (NaOH) to remove not only the cellulosic components but also the non-cellulosic components such as hemicellulose, lignin and pectin inside the plant fibre [1]. The alkali-treated fibres are then passed through multi-phase bleaching. Most of the manufacturers use this process as it requires not only a little time to yield the bamboo fibres but also less economic means mainly when compared with mechanical methods. Kumar et al. [47] in their study, soaked bamboo strips in 4% NaOH for 72 h to extract the fibre. This method removed 38–42% of the polysaccharides and lignin from the bamboo chips. The obtained pulp was cooled, filtered and washed, and then further treated with glacial acetic acid. Sodium chlorite was occasionally used to bleach the fibre to white. The treated pulp was called bleached bamboo fibre. The problem with this method was that fibre bundles with diameters of 100 ± 10.4 μm were also formed during the extraction; therefore, the parameters were chosen to optimise separation of bamboo fibre by using a minimum amount of NaOH [48].

In an exciting study, Kumar et al. [49] reported that the characteristic properties of mercerised bamboo fibres used for the preparation of bamboo fibre-reinforced epoxy composites made the bio composites cost useful for dielectric applications. In another interesting study reported by Kumar and Kumar [50], alkali treatment of bamboo fibre further increased the tensile and flexural strength of bamboo-epoxy nanocomposites by 60 and 42%, respectively, as compared to pure composites.

Many researchers have worked on the physical, mechanical and thermal behaviour (weathering behaviour, % water uptake, % thickness swelling and thermal stability), morphology properties and impact test of bamboo fibres reinforced novolac resin composites prepared using mercerised bamboo fibres. They reported that the modification improved various features such as fine structure, impact strength, wetting ability, interfacial strength, mechanical properties, weathering and thermal properties of the composites [51–53]. The effect of acrylic acid-grafted bamboo rayon on the antibacterial activity of acrylic acid-grafted bamboo rayon silver nanoparticles has been reported [54]. Bamboo rayon-copper nanoparticle composite fabric was also prepared using acrylic acid-grafted bamboo rayon revealed antibacterial activity against both Gram-positive and Gram-negative bacteria, which was durable until 50 washings [55].

In another study, the effects of alkaline and acetylating agents on the morphology of bamboo fibre-polypropylene were reported [56]. Their mechanical, thermal, rheological property, morphology and miscibility properties were extensively studied. The comparison of alkaline and acetylating treatments showed that the mechanical properties of bamboo fibre-polypropylene composites were improved and adhesion between bamboo fibre and polypropylene matrix was enhanced.

An HNO_3-$KClO_3$ method has also been used to extract fibre from bamboo samples. Before the addition of $KClO_3$, dry raw bamboo strands were immersed in a diluted nitric acid solution [57, 58]. After treating for 24 h at 50°C, the obtained bamboo fibre suspension was cooled then dialysed against distilled water to remove low molecular weight compounds. The slurry was then freeze-dried to obtain dry bamboo fibre.

He et al. [59] suggested a complicated method for obtaining bamboo fibre. Crude fibre bundles of bamboo, obtained by drawing bamboo chips roasted at 150°C for 30 min, were first immersed in water at 60°C for 24 h, and then air-dried before removing impurities further by repeated rolling. Subsequently, the fibre bundles were heated with 0.5% NaOH (w/v), 2% sodium sulphite, 2% sodium silicate and 2% sodium polyphosphate solutions for about 60 min at 100°C. The liquor-to-bamboo ratio was 20:1. After being washed with hot water, the fibres were treated with 0.04% xylanase and 0.5% DTPA (diethylene triamine pentaacetic acid) at 70°C for 60 min at a pH of 6.5. The fibres obtained were then heated for about 60 min at 100°C, except using 0.7% NaOH. In the bleaching step, the bamboo fibre was placed in a polyethylene bag with 4% H_2O_2, 0.2% NaOH and 0.5% sodium silicate for 50 min. The liquor ratio was 20, and the pH was maintained at 10.5. After treatment with 0.5% sulphuric acid solution for 10 min and then being emulsified for 5 days, refined bamboo fibre was obtained.

Maleic anhydride treatment has been reported to improve the mechanical (Modulus of elasticity and flexural modulus) as well as water-resistant properties (water uptake) of bamboo-epoxy composites [60]. The same trend in the properties of permanganate- and benzoylation-treated bamboo fibre polyester composites was observed [61]. In another study, the preparation of short bamboo fibre-reinforced polypropylene composites with various loadings percentages of chemically modified bamboo fibres was reported [62]. Maleic anhydride-grafted polypropylene was chosen, supported as a compatibiliser to improve adhesion between fibre and matrix.

Acrylonitrile treatment of bamboo fibre has been reported to improve the tensile, flexural and water absorption properties of acrylonitrile-treated bamboo fibre composites [49].

Anwar et al. [63] evaluated the effect of pressing time on physical and mechanical properties of phenolic-impregnated bamboo strips. The treatment of bamboo strips with low molecular weight phenol formaldehyde (LMwPF) resin followed by pressing at 140°C improved the dimensional stability and strength properties of the strips. The treatment improved water absorption, thickness swelling and linear expansion perpendicular to grain after 24 h of cold water soaking [63].

Liu et al. [64] developed and reported an efficient and eco-friendly technology for the improvement of interfacial adhesion of bamboo fibre-Unsaturated polyester (UPE) composites. A soybean oil-based monomer, acrylated epoxidised soybean oil (AESO) monomer was grafted onto the surface of bamboo fibre using 1,6-diisocyanatohexane (DIH) as a linker, and the attached C=C bonds participated in the crosslinking of UPE resins, thus forming chemical connections between the bamboo fibres and the UPE matrix. This resulted in significant improvements in both the static and dynamic mechanical properties of the composites by the improved interfacial adhesion. In addition, it significantly prevented the penetration and movement of water in the composites and resulted in reduced water uptake rates and diffusion coefficients.

Bamboo fibres could also be modified by atom transfer radical polymerisation (ATRP) technique. ATRP is a technique, which could allow the incorporation of polymers with predetermined molecular weight onto the surfaces of particles, fibres or membranes. It has been extensively studied for graft copolymerization of vinyl monomers onto fibres in a living/controllable way [65]. This method has made the incorporation of poly(poly(ethylene glycol) methyl ether methacrylate) (PPEGMA) onto the surfaces of bamboo fibres to improve interfacial adhesion between bamboo fibre and poly-(butylene succinate) [66].

Lu et al. [67] carried out a comparative study on the effect of alkali soaking, silane coupling agent and maleic anhydride grafting on the mechanical properties of cellulose/poly(L-lactic acid) composites. They reported that the modifications improved the mechanical properties of the cellulose/poly(L-lactic acid) composites by improving the interfacial adhesion of the cellulose fibre and the matrix. NaOH solution pre-treatment of cellulose provided the composites with the highest stiffness, KH560 modification resulted in best ductility. Maleic anhydride grafting onto poly(L-lactic acid) balanced the improvements of stiffness and ductility, exhibiting best overall properties [67].

Silanes are recognised as efficient coupling agents that are used extensively in composites and adhesive formulations. Silane coupling agents have a hydrophilic structure with different groups attached to the silicon atoms that can act as a bridge; one end is interacting with the matrix and the other end reacting with the hydrophilic fibres [64]. The silane coupling treatment has been found to significantly improve the impact fatigue strength of the composites [66]. A functional silane was employed as a coupling agent to enhance the interfacial adhesion between bamboo fibre and polylactic acid [68]. The silane-treated bamboo fibre-reinforced polylactic acid composite showed excellent mechanical and thermal properties compared with the properties of polylactic acid composite containing delignified BF. It was thought that effective surface treatment from the silane coupling agent-improved adhesion between the polylactic acid matrix and the bamboo fibre. In another study, Kushwaha and Kumar [69] investigated the effect of silanes on the mechanical properties of bamboo fibre-epoxy composites. They prepared two sets of bamboo-epoxy composites, one with silane treatment bamboo mats and the other with silane treatment mercerised bamboo mats. The mechanical properties (tensile strength, elastic modulus, flexural strength and flexural modulus) observed that silane treatment improved the tensile and flexural strength, but the addition of silane treatment mercerised bamboo leads to significant reduction of the strength. Similarly, Kumar and Kumar [50] in their work confirmed that the alkali and silane treatments with nano-clay filler improved the dielectric and mechanical properties of bamboo-epoxy composites.

Lee and Wang [70] used lysine-based diisocyanate (LDI) as a coupling agent for polylactic acid/bamboo fibre and poly(butylenes succinate)/bamboo fibre composites which improved their tensile and water resistance properties. Two novel bifunctional monomers, namely isocyanatoethyl-methacrylate and N-methylol acrylamide, have been used as the coupling agents to strength the interface of bamboo fibre/unsaturated polyester composites [71, 72].

A recently reported work reported an improved interfacial strength between poly(vinyl chloride) (PVC) and bamboo flour in PVC/bamboo flour composites using novel coupling agents [73]. One pot synthesis generated the coupling agents. The increased content of the coupling agents used increased the morphological and mechanical properties of composites. The result revealed that coupling agent enhanced the affinity between fibre and polyvinyl matrix by lowering down the interfacial tension. SEM studies carried out showed a better dispersion of fibre into the PVC matrix due to an increased amount of coupling agents used. The enhancement in mechanical properties was also an indication of strong bonding between matrix and bamboo fibre [73].

Dilute acid pre-treatment of bamboo shoots shell fibre (BSSF) and bamboo stem and leaf (BSL) have been investigated for xylose and glucose yields [42]. Pre-treatment of bamboo (*Dendrocalamus asper*) with dilute sulphuric acid before enzymatic hydrolysis process to produce fermentable sugars has also been investigated [38]. Dilute phosphoric acid pre-treatment of bamboo was also studied for producing dissolving pulp for textile utilisation [74].

Recently in another study, a solvent (concentrated phosphoric acid) and organic solvent (95% ethanol)-based lignocellulose fractionation (COSLIF) methods have been developed to pretreat bamboo [34].

4. Utilisation of bamboo resources

Bamboo plays a significant role as a material for consumer products. With its high growth rate, a wide range of applications and renewability, bamboo resources occupy a noteworthy position in the twenty-first century as a versatile and vital raw material [20]. Besides, it is recognised as an industrial raw material globally and has tremendous potentials for economic development of nations [75]. The bamboo potential as an industrial raw material is linked to its agronomical and technological characteristics [2]. Given its proprieties, bamboo continues to be used for the production of new products. The multi-functional ranges of bamboo uses have shown that it may prove beneficial as a valuable and sustainable natural resource [76]. In China, it is the valuable raw material for the booming bamboo industry. Bamboo has been known to find large applications for both food and non-food industries. At present, there are about 3000 companies around the world that are engaged in the production of various bamboo-based products [77]. Bamboo provides food, shelter and medicine and serves as raw material for many industries. It has found application in different industrial sectors including civil construction, wood, paper and pulp, textile, electrical and electronics, agriculture and agro-allied, food, chemical and pharmaceutical, reinforcement, automobile and medicine. Some bamboo-based products include house construction materials, household items, biofuel, chemical and pharmaceutical products, pulp and paper, irrigation and drainage pipes and textiles materials, panels, flooring materials, charcoal, edible shoots and other daily-use articles [75, 77]. The use of bamboo fibres as reinforcement in composite materials has been immensely amplified with high-tech revolution in recent years. This is as a response to the increasing demand for developing materials that are biodegradable, sustainable and recyclable [15]. Apart from the above, several other applications exist, though on a relatively small scale. For instance, there are some situations where bamboo is used as poles for aerial antenna, electrification, rafters, fishing traps, yam stakes, etc. Ladapo et al. [75] further reported that the current uses of bamboo in Nigeria represent only a fraction of economic activities in the country.

Besides its role as a raw material for consumer products, bamboo has enormous prospects for industrial utilisation and as industrial raw material. Industrial application of bamboo is in the areas highlighted below:

4.1. Bamboo human food products

Bamboo is a good source of food particularly the shoot. It is a delicacy in Asia. In India, young and tender bamboo shoots are used as a seasonal vegetable in both rural and urban areas [78]. It has been reported to include natural products, such as potassium, carbohydrates, dietary fibres, vitamins and other active materials, which are used for traditional food in many countries [79] and further conversion of these carbohydrates, give rise to other products like xylitol. Figures of nutrient contents of *Bambusa vulgaris* show it to contain crude protein (10.1 g), crude fibre (21.7 g), ether extract (2.5 g), ash (21.3 g), phosphorous (86 mg), iron (13.4 mg), vitamin B1 (0.1 mg), vitamins B2 (2.54 mg) and carotene (12.3 mg)/100 g).

Tea made from the bamboo leaf is rich in silica, which is important in bone and other rigid tissue health. Silica improves bone health, strengthens hair and nails, improves dental health and make the skin more elastic and healthy [80]. Bamboo leaf tea is a low-calorie health food, which is rich in protein and fibre, but free of caffeine. As many cups as possible can be taken as bamboo tea stimulates metabolism without side effects [3]. The common species of bamboo for this purpose are given in **Table 5**.

4.2. Bamboo charcoal

Bamboo charcoal can be used in different industries including chemical, pharmaceutical and energy production industries (**Table 5**). Recent studies have shown that bamboo species are also a good source of quality activated carbon, which can find application in medicine, foods, chemical and metallurgical industries. Activated bamboo charcoal has found application in cleaning the environment, absorbing excess moisture and producing medicines [81]. Bamboo charcoal is generally used by goldsmith and in gardening to prevent moisture available to plants particularly in Japan [78].

Bamboo charcoals are multi-functional materials pyrolysed from bamboo under anaerobic conditions. During this pyrolytic process, bamboo is converted to stable charcoal. It serves as a substitute for wood charcoal or mineral coal and has been reported to possess absorption capacity which is six times that of wood charcoal of the same weight [3, 14]. Hence, it is a suitable absorbent. Studies have also revealed their uses as absorbent for dyes [82–85], heavy metal [86, 87], organic pollutants [88] and other substances [89, 90]. Other applications include for purification of waters, soils and sediments contaminated by PAHs; for environmental protection and architectural decorations [91] and as conductor and fuel [14].

Industry	Part of bamboo	Bamboo species	Bamboo products	References
Pulp and paper	Culm	*Bambusa vulgaris*	Pulp, paper, paper board	Okumura et al. [2], Ogunwusi and Onwualu [3]
Energy production	Stem, branch and rhizome	*Dendrocalamus sp., Bambusa vulgaris*	Bamboo charcoal, bamboo charcoal briquettes, bio-methane, Bio-ethanol	Ogunwusi and Onwualu [3], He et al. [7], Biopact [93], Kuttiraja et al. [35];Okumura et al. [2], Azzini et al. [96]
Wood	Culm	*Phyllostachys aurea, Guadua angustifolia, Melocanna baccifera*	Plywood, laminated bamboo board, mat ply bamboo, curtains ply bamboo, laminated wood strips, mat curtain plywood, door shutter, matchstick, bamboo clipboard, furniture and handicrafts, charcoal	Okumura et al. [2], Xuhe [77], Ogunwusi and Onwualu [3], Ogunwusi [105]
Chemicals and pharmaceuticals	Leaves, shoot, culm	*Bambusa vulgaris*	Cellulose, bio-ethanol, bio-methane, starch, charcoal, flavour and preservatives, bamboo leaf tea	Okumura et al. [2], He et al. [7], Costa et al. [43], Azzini and Gondim-Tomaz [106], Azzini et al. [96], Ogunwusi and Onwualu [3]
Reinforcement	Culm	*Guadua augustifolia, G. brang, G. levis, G. scortechini, G. wrayi*	Floor tiles, toys, bamboo composite decking, bamboo composite fencing, bamboo composite deck tiles, bamboo composite railings, bamboo composite dustbins, bamboo composite outdoor furniture, bamboo decking accessories	Khalil et al. [15], Lee and Wang [70], Ogunwusi [105]
Civil construction	Culm	*Bambusa balcooa, B. tulda, B. nutans, B. pallida, B, polymorpha, Dendrocalamus hamiltonii, Melocanna baccifera, D. giganteus, Gigantochloa apus, Guadua augustifolia*	Irrigation and drainage pipes, flooring & floor covering, lamination of strip (plywood), decking, bamboo scaffolding and cladding, engineered bamboo flooring products, high-density beams and panels, composite board, particle board, bamboo prefabricated house.	Ladapo et al. [75], Lobovikov et al. [14], Ogunwusi and Onwualu [3], Lipangile [102], Akinbile et al. [101]

Industry	Part of bamboo	Bamboo species	Bamboo products	References
Food and cottage	Shoot, culm	*D. latiflorus, Dendrocalamus asper, D. hamiltonii, D. brandisii, D. strictus, D. latifloru, Bambusa blumenn,B. tulda, B. nutans, B. balcooa, B. polymorpha, B. pallida, Thyrsostachys oliveri, Gigantochloa albociliata, Melocanna baccifera*	Bamboo vegetable, bamboo edible shoots, bakery products, diary products, meat and aquatic products	Miura et al. [107], Okumura et al. [2], Xuhe [77], Parajó et al. [108], Nabarlatz et al. [109], Vazquez et al. [110], Aoyama and Seki [111], Aoyama and Seki [112], Chongtham et al. [113], Borah et al. [78]
Textile	Leaves, woody shoot	*Phyllostachys edulis and Bambusa emeiensis*	Bamboo cloth, bamboo rayon, bamboo linen	Teli and Sheikh [54], Ogunwusi and Onwualu [3], Waite and Platts [114]
Electrical and electronics	Culm	*Guadua augustifolia, G. brang, G. levis, G. scortechini, G. wrayi*	Hardware for electronics, wrist watches, chains, fan blades	Lee and Wang [70]
Agricultural and agro-allied	Leaves	*Yushania alpina* and *Oxytenanthera abyssinica*	Ornamental plants, bamboo livestock feeds	Mekuriaw et al. [115]
Automobile	Culm	*Guadua augustifolia*	Prototype of door trim	Khalil et al. [15]
Medicine	Leaves, culm, stem, sap	*Phyllostachys Sieb. et Zucc.* family (especially *P. nigra* var. *henonis*, *P. edulis* and *P. bambusoides, Bambusa bambos, Pleioblastus amarus, B. spinosa*	Antioxidants, flavonoids	Hu et al. [116], Jiao et al. [117], Xie et al. [118]

Table 5. Bamboo industries, parts, species and products.

However, bamboo is a green biofuel for fighting deforestation and climate change [92]. Because of its excellent characteristics, some countries are renewing their bioenergy strategies to include bamboo. It is worthy of note that a bamboo fuelled power station is being built in Mizoram state of India to help meet the energy need of India's northeast [93]. Bamboo may, therefore, serve as the key to combating energy problem, especially in developing nations if adopted. Common bamboo species that can be used for this purpose (see **Table 5**).

4.3. Bamboo cellulose and alcohol

Bamboos have agronomical and technological characteristics highly essential to obtain cellulose [2]. Their potential and utilisation for cellulose, bio-ethanol and other related products have been studied. It has been recognised as a useful resource for this purpose due to its higher growth rate, annual biomass yield and significant amount of sugars [2, 21]. He et al. [21] reported that bio-ethanol production from bamboo can also follow the general process for ethanol production from other lignocellulosic materials, which including pre-treatment, enzymatic hydrolysis and fermentation. The high contents of starch and cellulose, 70–85% of the stem, have given bamboo an advantage as an alternative source of alcohol production. The part transformed to alcohol represents the yield of 250–380 l ton^{-1} of bamboo [94] while that of sugar cane, when compared gave an average yield of 70 l of alcohol ton^{-1} [95]. In another study, Azzini et al. [96] developed a combined production process for cellulose fibres and ethanol from *B. vulgaris, which* gave good results to bamboo use. Other common species of bamboo used for this purpose are shown in **Table 5**.

4.4. Bamboo pulp and paper

Bamboo is endowed with a long fibre length, and this makes bamboo pulp suitable for paper-making. It has been found out that from 4 tons of bamboo nearly 1 ton of pulp is produced which is utilised in different furnishes for production of paper and board [78]. China and India use bamboo mainly for producing pulp and paper. Bamboo paper has the same quality with paper made from wood. Its brightness and optical properties remain stable while those papers make from wood may deteriorate over time. The morphological characteristics of bamboo fibres give paper made from bamboo a high tear index [97].

4.5. Bamboo medicinal products

Bamboo plant is considered medicinal. It has been found to have a high level of acetylcholine, which acts as a neurotransmitter in animals and humans [98]. Culms of many bamboo species secret siliceous materials, which can be used for medicine. This siliceous material is used as a cooling tonic and aphrodisiac, as a remedy for asthma and coughs [98] and other debilitating diseases. Medicine made from the leaves of *Pleioblastus amarus* is used for treating fever, fidgeting and lungs inflammation [98]. Stems and leaves of *Bambusa bambos* are used in traditional Indian medicine as a blood purifier, in the treatment of leucoderma and inflammatory conditions. It is also given internally for treatment of bronchitis, gonorrhoea and fever. The burnt roots of this species are used to treat ringworm, bleed gums and painful joints. The bark is a cure for eruptions [99]. The leaf bud of *B. spinosa* is used in leprosy, fever and haemoptysis. The sap of *B. vulgaris* is given as a remedy for phthisis in the Philippines [100]. Extractives from various parts of the bamboo plant have been used for hair and skin ointment, medicine for asthma, eyewash, potions for lovers and poison for rivals. Bamboo shoot is one vegetable that is free in pollution, low in fat and high in edible fibre and rich in mineral. It functions well in removing sputum, enhancing digestion, relieving toxicity, improving dieresis and it is frequently used for healing swollen tissues or oedema and abdominal disease in which watery fluids collect in cavities or body tissues [76]. The shoot also contains saccharine, which can resist little mouse tumour and has anti-ageing elements [20, 80].

4.6. Bamboo culm drainage pipes

Drainpipes made from bamboo serve as low-cost substitute to those made from assorted materials. The use of drainage pipes made of various assorted materials is very common in our markets [101]. These materials are very expensive, not readily available, require a high degree of maintenance and pollute water, which they convey due to the pipe's constituents. They include drainpipes made from wood boards or box drains, bricks, horseshoe-shaped ceramic tile, circular clay tile, concrete tile, bituminised fibre perforated pipe, perforated smooth plastic pipe to corrugated plastic pipe. Currently, corrugated pipes are frequently used, although clay and concrete pipes are still being used as well. In rural Tanzania, a bamboo pipe network is being used for providing safe and constant water supply to a large rural population [102]. In a study reported by Akinbile et al. [101] found that the use of bamboo (*B. vulgaris*) of predetermined lengths and diameters as a field drainage material was found effective. The result indicated that bamboo could be satisfactorily used as an alternative to the various assorted materials that are very common in markets to provide an advantage of cost, as well as easy transportation, handling and laying. Also, bamboo pipes do not contaminate the water being conveyed and do not react with the soil; unlike the other assorted materials, thus preventing the excessive cost of treating the water being conveyed for the various human and animal uses [101].

5. Bamboo building or civil construction materials

Bamboo is a principal construction material in many countries, particularly in rural areas. It can be used for almost all parts of houses, including posts, roofs, walls, floors, beams, trusses and fences [14]. At present, the bamboo utilisation in construction is of subjective form and based generally in traditional systems established in each country. In Asia, bamboo is quite common for bridges, scaffolding and housing, but it is usually a temporary exterior structural material. In many overpopulated regions of the tropics, certain bamboos supply the one suitable material that is sufficiently cheap and plentiful to meet the great need for economical housing [103]. However, bamboo offers an ecologically viable alternative to timber for construction due to its low-cost and fast growth rate [3]. Prior to bamboo utilisation in large scale as an economically viable engineering material, a study on properties and structures must be carried out. These studies can provide better cultivation, harvest and processing techniques, besides analysis of mechanical and physical properties of bamboo stems. Bamboo stems present excellent physical and mechanical properties that can be used as a substitute for other materials such as steel aiming fabrication of concrete structures. Several studies were conducted on physical and mechanical characteristics to test the bamboo as construction material. The mechanical properties of bamboo are strongly affected by the age, species and humidity. Bamboos do not break easily, and its original shape is regained when the load subjected to it is removed. It is suitable for reinforcing concrete [75]. Some bamboo products for building or construction purpose include ply bamboo, bamboo panels/composite boards of different types, particle boards, mat boards, bamboo parquet and bamboo fibre-reinforced plastic [5, 76, 104]. The common species of bamboo used for this purpose are given in **Table 5**.

6. Bamboo livestock feed and ornamental plant

The Japanese have used the leaves of bamboo as fodder for livestock for hundreds of years. It is also an essential food for the giant pandas in China because they survive only on bamboos. Many bamboos are popularly used as ornamental plants to beautify homes and gardens. Ogunwusi and Onwualu [3] reported that feeding chickens on organic diets containing fresh bamboo leaves lead to 70% weight gain more than those do fed on standard organic diets. This suggests that the fibre in the bamboo leaves enlarge the digestive tract and enable the chicken to consume more and grow faster.

7. Bamboo chemicals and pharmaceutical products

The ashes of bamboo are used to polish jewels and manufacture electrical batteries. Bamboo beverage and beer have been widely accepted mainly in Asian countries such as China, Korea and Japan. Bamboo has been reported to be valuable in health care delivery and processed into beverages, medicines, pesticides and other household items such as toothpaste, soaps, etc. The culm, shoot and leaf of bamboo have been reported to possess anti-oxidation, anti-ageing, antibacterial and antiviral properties. Some materials extracted from bamboo are used in fresh flavour and preservation of food. This is due to the nutritious and active minerals (such as vitamins, amino acids, flavine, phenolic acids, polysaccharide, trace elements and steroids) they contain [76]. Bamboo leaf contains 2–5% flavine and phenolic compound that have the power to remove active oxy-free radicals, stopping nitrification and abating blood fat. Of late, research has shown bamboo charcoal as one of the base materials for human health, from water treatment to its uses as a shield from electromagnetic radiation. Bamboo extracts contain valuable elements, which are used in pharmaceuticals, creams and beverages. Bamboo gas can be used as a substitute for petroleum. Activated charcoal is used as a deodorant, purifier, disinfectant, medicine, agricultural chemicals and an absorbent of pollution and excessive moisture. A number of researchers have also produced bamboo ethanol and butanol. Butanol is similar to ethanol being a form of alcohol. It has high-density; non-corrosive and can, therefore, be mixed with gasoline. Another significant product from bamboo is activated carbon. Activated carbon is a non-graphitic form of carbon, which is produced by activation of any carbonaceous material such as bamboo, wood chips, etc. Activated carbon is employed in the decolourisation of sugar and sweeteners, drinking water treatment, gold recovery, production of pharmaceuticals and fine chemicals, catalytic process, off-gas treatment of waste incinerators, automotive vapour filters, colour/odour correction in wines and fruit juices.

Some chemical and pharmaceutical products are cellulose, bio-ethanol, bio-methane, starch, charcoal, flavour and preservatives, bamboo leaf tea (see **Table 5**).

8. Bamboo handicraft products

Bamboo is known as one of the materials more versatile to the handicraft production because it is a raw material of easy acquisition, low-cost and demands simple tools in preparation. Besides, bamboo is also a material of high plasticity and easy to combine with other materials. Hence, it allows an artist to express his talent in several forms. Handicraft products with bamboo are exported mainly from Asian countries to other parts of the world. The excellent quality of these products is due to good qualification of the workforce and the availability of machines that assist the production process.

Bamboos are used for making skewers, chopsticks, boats, weapons, matchsticks, containers, poles, bows and arrows, rafts and fishing poles. Various craft products made of bamboo are baskets, tools, handles, hats, traditional toys, mat, flooring material, purses, bags, satchels, tea packaging, floor tiles, general household product, furniture, utensils, musical instruments, etc. Craft products are to be developed differently, marketed and promoted in innovative and various ways if they are to compete and survive in the international market. **Table 5** shows some typical bamboo species and parts used for this purpose [78].

9. Bamboo furniture

Bamboo serves as an excellent alternative to wood in the manufacture of furniture. It competes in market of wood without restrictions. In Asia, there is a significant production of furniture of bamboo. Artists of this continent have refined techniques that qualify these products to exigent markets. Various bamboo furniture products including bamboo panel, composite board and particleboard have been developed in China [76]. Bamboo mat board is being manufactured in China, India, Thailand and Vietnam [5]. Other bamboo furniture products are beds, cupboard, table, upholstery chairs [75], ply bamboo, laminated bamboo, mat ply bamboo, curtains ply bamboo, laminated bamboo strips, mat curtain plywood, bamboo chipboard, floor tiles and composites. Laminated bamboo furniture is on the rise rapidly in the world especially Asia. When bamboo is laminated and used to produce furniture, it is difficult to differentiate it from wood [75]. **Table 5** shows some typical bamboo species and parts used for this purpose.

10. Conclusion

The application of bamboo in many spheres of life as a suitable replacement for woody biomass in some instances provides arrays of opportunities in material industries. Bamboo being trees that grow and mature in a very short period of time can secure continuous supply of cheap tree and fibres compared to other woody biomasses. In addition, the excellent properties of bamboo enhanced via different chemical modification methods to improve their mechanical and thermal makes their fibre to be competitive with other materials used in the reinforcing of different polymers. With the application of correct treatment, suitable fibres can be extracted from bamboo for various purposes. And as such, several outstanding bamboo fibres can be obtained and incorporate other materials to produce excellent performance composites that can favourably compete with many conventional materials. The use of alkali hydrolysis has been majorly to remove the amorphous regions that are responsible for low resistance to fungus attack in the fibre. The interfacial adhesion between the fibre and other matrices can enhance the inclusion of coupling agents and fibre pre-treatment using acid hydrolysis. The use of bamboo fibres as a replacement for petroleum-derived fibres can breed the development of eco-friendly products that exert less pressure on the environment.

Bamboo fibres have great potential as an alternative to inorganic fillers and are raw material for fabricating a composite material, and their applicability is being widely investigated. Green composites, defined as biodegradable biopolymers reinforced by natural fibres, have very low impact on the environment, and thus they are one of the potential alternatives to replace conventional petroleum-based polymers and polymer composites. Bamboo has been used widely utilised in pulp and paper, energy production, chemicals and pharmaceuticals, reinforcement, civil construction, textile, electrical and electronics, agricultural and agro-allied, automobile and medicine. Although some of these applications are very limited in scope, continuous researches on bamboo can increase their potential and induce their aggressive interest in many more areas.

Author details

Mayowa Akeem Azeez* and Joshua Iseoluwa Orege

*Address all correspondence to: mayowa.azeez@eksu.edu.ng

Department of Industrial Chemistry, Ekiti State University, Ado Ekiti, Nigeria

References

[1] Liu D, Song J, Anderson DP, Chang PR, Hua Y. Bamboo fiber and its reinforced composites: Structure and properties. Cellulose. 2012;**19**(5):1449-1480

[2] Okumura RS, Queiroz R, Takahashi LSA, Santos D, Lobato A, Mariano D d C, Aves G, Fillo BB. Plant morphology, agronomic aspects, human utilization and perspectives. Journal of Food, Agriculture and Environment. 2011;**9**(2):778-782

[3] Ogunwusi A, Onwualu A. Prospects for multi-functional utilisation of bamboo in Nigeria. Prospects. 2013;**3**:8

[4] Patel Pratima A, Maiwala Adit R, Gajera Vivek J, Patel Jaymin A, Magdallawala Sunny H. Performance evaluation of bamboo as reinforcement in design of construction element. International Refereed Journal of Engineering and Science. 2013;**2**(4):55-63

[5] Sattar M, Kabir M, Bhattacharjee D. Effect of age and height position of muli (*Melocanna baccifera*) and borak (*Bambusa balcooa*) bamboos on their physical and mechanical properties. Bangladesh. Journal of Forest Science. 1990;**19**(1/2):29-38

[6] Kabir M, Bhattacharjee D, Sattar M. Effect of age and height on strength properties of Dendrocalamus longispathus. Bamboo Information Centre India Bulletin. 1993;**3**(1):11-15

[7] He M-X, Wang J-L, Qin H, Shui Z-X, Zhu Q-L, Wu B, Tan F-R, Pan K, Hu Q-C, Dai L-C. Bamboo: A new source of carbohydrate for biorefinery. Carbohydrate Polymers. 2014;**111**:645-654

[8] Kleinhenz V, Midmore DJ. Aspect of Bamboo Agronomy. City, 2001

[9] Scurlock J, Dayton D, Hames B. Bamboo: An overlooked biomass resource? Biomass and Bioenergy. 2000;**19**(4):229-244

[10] MCI. Millennium Investment Initiative. Investment Opportunity in Kumasi, Ghana. Bamboo cultivation and processing. City. 2013. Available at: http://mci.ei.columbia.edu/files/2013/10/Kumasi-Bamboo-Cultivation-and-Processing.pdf. [Accessed 20th September, 2017]

[11] Pérez MR, Maogong Z, Belcher B, Chen X, Maoyi F, Jinzhong X. The role of bamboo plantations in rural development: The case of Anji County, Zhejiang, China. World Development. 1999;**27**(1):101-114

[12] Dransfield S, Widjaja E. Plant Resources of South-East Asia. Pudoc; 1995

[13] Maoyi F, Baniak R. Bamboo Production Systems and their Management. Proceedings of the Vth International Bamboo Workshop and the IV International Bamboo Congress on Bamboo, People and Environment. Ubud, Bali, Indonesia; June 1995:19-22

[14] Lobovikov M, Ball L, Guardia M, Russo L. World Bamboo Resources: A Thematic Study Prepared in the Framework of the Global Forest Resources Assessment 2005. Rome: Food & Agriculture Org.; 2007

[15] Khalil HA, Bhat I, Jawaid M, Zaidon A, Hermawan D, Hadi Y. Bamboo fibre reinforced biocomposites: A review. Materials & Design. 2012;**42**:353-368

[16] Kassahun T. Review of bamboo value chain in Ethiopia. International Journal of African Society Culture and Traditions. 2014;**2**(3):52-67

[17] Fu MY, Banik RL. Bamboo Production Systems and their Management. City, 1995

[18] Kelbessa E, Bekele T, Gebrehiwot A, Hadera G. A Socio-Economic Case Study of the Bamboo Sector in Ethiopia: An Analysis of the Production-to-Consumption System. Ethiopia: Addis Ababa; 2000

[19] Tinsley BL. Bamboo Harvesting for Household Income Generation in the Ethiopian Highlands: Current Conditions and Management Challenges. Graduate Student Theses, Dissertations, & Professional Papers. 2015. https://scholarworks.umt.edu/etd/4369

[20] Salam K. Bamboo for economic prosperity and ecological security with special reference to north-East India, CBTC, Guwahati. Indian folklore Journals. 2008;**409**:353

[21] He M-X, Wang J-L, Qin H, Shui Z-X, Zhu Q-L, Wu B, Tan F-R, Pan K, Hu Q-C, Dai L-C, Wang W-G, Tang X-Y, Hu GQ. Bamboo: A new source of carbohydrate for biorefinery. Carbohydrate Polymers. Oct 13, 2014;**111**:645-654

[22] Kobayashi F, Take H, Asada C, Nakamura Y. Methane production from steam-exploded bamboo. Journal of Bioscience and Bioengineering. Jan 1, 2004;**97**(6):426-428

[23] Mishra S, Mohanty AK, Drzal LT, Misra M, Hinrichsen G. A review on pineapple leaf Fibers, sisal Fibers and their biocomposites. Macromolecular Materials and Engineering. 2004;**289**(11):955-974

[24] Vogtlander J, Van der Lugt P. The environmental impact of industrial bamboo products: Life-cycle assessment and carbon sequestration. Techical Report No. 35. International Network for Bamboo and Ratton. 2015

[25] Kumar S, Dobriyal P. Treatability and flow path studies in bamboo. Part I. *Dendrocalamus strictus* needs. Wood and Fiber Science. 1992;**24**(2):113-117

[26] Chen X, Zhang X, Zhang Y, Booth T, He X. Changes of carbon stocks in bamboo stands in China during 100 years. Forest Ecology and Management. 2009;**258**(7):1489-1496

[27] Zhou G, Meng C, Jiang P, Xu Q. Review of carbon fixation in bamboo forests in China. The Botanical Review. 2011;**77**(3):262

[28] Ben-Zhi Z, Mao-Yi F, Jin-Zhong X, Xiao-Sheng Y, Zheng-Cai L. Ecological functions of bamboo forest: Research and application. Journal of Forestry Research. 2005;**16**(2):143-147

[29] Nath AJ, Das G, Das AK. Above ground biomass, production and carbon sequestration in farmer managed village bamboo grove in Assam, northeast India. Bamboo Science & Culture. 2008;**21**:1

[30] Hong C, Fang J, Jin A, Cai J, Guo H, Ren J, Shao Q, Zheng B. Comparative growth, biomass production and fuel properties among different perennial plants, bamboo and miscanthus. The Botanical Review. 2011;**77**(3):197

[31] Li Q, Song J, Peng S, Wang JP, Qu G-Z, Sederoff RR, Chiang VL. Plant biotechnology for lignocellulosic biofuel production. Plant Biotechnology Journal. 2014;**12**(9):1174-1192

[32] Nurul Fazita M, Jayaraman K, Bhattacharyya D, Mohamad Haafiz M, Saurabh CK, Hussin MH, HPS AK. Green composites made of bamboo fabric and poly(lactic) acid for packaging applications—A review. Materials. 2016;**9**(6):435

[33] Li L-J, Wang Y-P, Wang G, Cheng H-T, Han X-J. Evaluation of properties of natural bamboo fiber for application in summer textiles. Journal of Fiber Bioengineering and Informatics. 2010;**3**(2):94-99

[34] Sathitsuksanoh N, Zhu Z, Ho T-J, Bai M-D, Zhang Y-HP. Bamboo saccharification through cellulose solvent-based biomass pretreatment followed by enzymatic hydrolysis at ultra-low cellulase loadings. Bioresource Technology. 2010;**101**(13):4926-4929

[35] Kuttiraja M, Sindhu R, Varghese PE, Sandhya SV, Binod P, Vani S, Pandey A, Sukumaran RK. Bioethanol production from bamboo (Dendrocalamus sp.) process waste. Biomass and Bioenergy. 2013;**59**:142-150

[36] Yasuda M, Takeo K, Matsumoto T, Shiragami T, Sugamoto K, Matsushita Y-I, Ishii Y. Effectiveness of Lignin-Removal in Simultaneous Saccharification and Fermentation for Ethanol Production from Napiergrass, Rice Straw, Silvergrass, and Bamboo with Different Lignin-Contents. Rijeka, Croatia: InTech; 2013:91-104

[37] Yamashita Y, Shono M, Sasaki C, Nakamura Y. Alkaline peroxide pretreatment for efficient enzymatic saccharification of bamboo. Carbohydrate Polymers. 2010;**79**(4):914-920

[38] Leenakul W, Tippayawong N. Dilute acid pretreatment of bamboo for fermentable sugar production. Journal of Sustainable Energy & Environment. 2013;**1**(3):117-120

[39] Li Z, Jiang Z, Fei B, Cai Z, Pan X. Comparison of bamboo green, timber and yellow in sulfite, sulfuric acid and sodium hydroxide pretreatments for enzymatic saccharification. Bioresource Technology. Jan 1, 2014;**151**:91-99

[40] Littlewood J, Wang L, Turnbull C, Murphy RJ. Techno-economic potential of bioethanol from bamboo in China. Biotechnology for Biofuels. 2013;**6**(1):173

[41] Li Y, Zhang J, Chang SX, Jiang P, Zhou G, Fu S, Yan E, Wu J, Lin L. Long-term intensive management effects on soil organic carbon pools and chemical composition in Moso bamboo (*Phyllostachys pubescens*) forests in subtropical China. Forest Ecology and Management. 2013;**303**:121-130

[42] He M, Li Q, Liu X, Hu Q, Hu G, Pan K, Zhu Q, Wu J. Bio-ethanol production from bamboo residues with lignocellulose fractionation technology (LFT) and separate hydrolysis fermentation (SHF) by *Zymomonas mobilis*. American Journal of Biomass and Bioenergy. 2013;**1**:1-10

[43] Costa MME, Melo SLS, Santos JVM, Arujo EA, Cunha GP, Deus EP, Schmitt N. Influence of physical and chemical treatments on the mechanical properties of bamboo fibers. Procedia Engineering. 2017;**200**:457-464

[44] Fei M-E, Xie T, Liu W, Chen H, Qiu R. Surface grafting of bamboo fibers with 1, 2-epoxy-4-vinylcyclohexane for reinforcing unsaturated polyester. Cellulose. 2017;**24**(12):5505-5514

[45] Khoo S, Johar M, Low K, Wong K. Interfacial shear strength characterisation of alkali treated bamboo bundle-polyester composites using an improved technique. Plastics, Rubber and Composites. 2017;**46**(10):450-457

[46] Varshney V, Naithani S. Chemical Functionalization of Cellulose Derived from Nonconventional Sources. City: Springer; 2011

[47] Kumar S, Choudhary V, Kumar R. Study on the compatibility of unbleached and bleached bamboo-fiber with LLDPE matrix. Journal of Thermal Analysis and Calorimetry. 2010; **102**(2):751-761

[48] Deshpande AP, Bhaskar Rao M, Lakshmana Rao C. Extraction of bamboo fibers and their use as reinforcement in polymeric composites. Journal of Applied Polymer Science. 2000;**76**(1):83-92

[49] Kumar V, Kushwaha PK, Kumar R. Impedance-spectroscopy analysis of oriented and mercerized bamboo fiber-reinforced epoxy composite. Journal of Materials Science. 2011; **46**(10):3445-3451

[50] Kumar V, Kumar R. Dielectric and mechanical properties of alkali-and silane-treated bamboo-epoxy nanocomposites. Journal of Composite Materials. 2012;**46**(24):3089-3101

[51] Das M, Prasad V, Chakrabarty D. Thermogravimetric and weathering study of novolac resin composites reinforced with mercerized bamboo fiber. Polymer Composites. 2009;**30**(10):1408-1416

[52] Das M, Chakraborty D. Processing of the Uni-directional powdered phenolic resin—Bamboo fiber composites and resulting dynamic mechanical properties. Journal of Reinforced Plastics and Composites. 2009;**28**(11):1339-1348

[53] Rajulu AV, Baksh SA, Reddy GR, Chary KN. Chemical resistance and tensile properties of short bamboo fiber reinforced epoxy composites. Journal of Reinforced Plastics and Composites. 1998;**17**(17):1507-1511

[54] Teli M, Sheikh J. Antibacterial and acid and cationic dyeable bamboo cellulose (rayon) fabric on grafting. Carbohydrate Polymers. 2012;**88**(4):1281-1287

[55] Teli M, Sheikh J. Bamboo rayon–copper nanoparticle composites as durable antibacterial textile materials. Composite Interfaces. 2014;**21**(2):161-171

[56] Phuong NT, Sollogoub C, Guinault A. Relationship between fiber chemical treatment and properties of recycled pp/bamboo fiber composites. Journal of Reinforced Plastics and Composites. 2010;**29**(21):3244-3256

[57] Liu D, Zhong T, Chang PR, Li K, Wu Q. Starch composites reinforced by bamboo cellulosic crystals. Bioresource Technology. 2010;**101**(7):2529-2536

[58] Liu H, Huang Y, Yuan L, He P, Cai Z, Shen Y, Xu Y, Yu Y, Xiong H. Isothermal crystallization kinetics of modified bamboo cellulose/PCL composites. Carbohydrate Polymers. 2010;**79**(3):513-519

[59] He JX, Tang Y, Wang SY. Differences in morphological characteristics of bamboo fibres and other natural cellulose fibres: Studies on X-ray diffraction, solid state 13c-cp/mas nmr, and second derivative ftir spectroscopy data. Iranian Polymer Journal. 2007;**16**(12):807-818

[60] Kushwaha PK, Kumar R. Studies on performance of acrylonitrile-pretreated bamboo-reinforced thermosetting resin composites. Journal of Reinforced Plastics and Composites. 2010;**29**(9):1347-1352

[61] Kushwaha PK, Kumar R. Influence of chemical treatments on the mechanical and water absorption properties of bamboo fiber composites. Journal of Reinforced Plastics and Composites. 2011;**30**(1):73-85

[62] Chattopadhyay SK, Khandal R, Uppaluri R, Ghoshal AK. Bamboo fiber reinforced polypropylene composites and their mechanical, thermal, and morphological properties. Journal of Applied Polymer Science. 2011;**119**(3):1619-1626

[63] Anwar U, Paridah M, Hamdan H, Sapuan SM, Bakar E. Effect of curing time on physical and mechanical properties of phenolic-treated bamboo strips. Industrial Crops and Products. 2009;**29**(1):214-219

[64] Liu W, Xie T, Qiu R. Bamboo fibers grafted with a soybean-oil-based monomer for its unsaturated polyester composites. Cellulose. 2016;**23**(4):2501-2513

[65] Carlmark A, Malmström E. Atom transfer radical polymerization from cellulose Fibers at ambient temperature. Journal of the American Chemical Society. Feb 1, 2002;**124**(6):900-901

[66] Bao L, Chen Y, Zhou W, Wu Y, Huang Y. Bamboo fibers@ poly(ethylene glycol)-reinforced poly(butylene succinate) biocomposites. Journal of Applied Polymer Science. 2011;**122**(4): 2456-2466

[67] Lu T, Liu S, Jiang M, Xu X, Wang Y, Wang Z, Gou J, Hui D, Zhou Z. Effects of modifications of bamboo cellulose fibers on the improved mechanical properties of cellulose reinforced poly(lactic acid) composites. Composites Part B: Engineering. 2014;**62**:191-197

[68] Kang JT, Kim SH. Improvement in the mechanical properties of polylactide and bamboo fiber biocomposites by fiber surface modification. Macromolecular Research. 2011; **19**(8):789-796

[69] Kushwaha PK, Kumar R. Effect of silanes on mechanical properties of bamboo fiber-epoxy composites. Journal of Reinforced Plastics and Composites. 2010;**29**(5):718-724

[70] Lee S-H, Wang S. Biodegradable polymers/bamboo fiber biocomposite with bio-based coupling agent. Composites Part A: Applied Science and Manufacturing. 2006;**37**(1):80-91

[71] Liu K, Takagi H, Osugi R, Yang Z. Effect of lumen size on the effective transverse thermal conductivity of unidirectional natural fiber composites. Composites Science and Technology. 2012;**72**(5):633-639

[72] Liu L, Yu Y, Yan C, Li K, Zheng Z. Wearable energy-dense and power-dense supercapacitor yarns enabled by scalable graphene–metallic textile composite electrodes. Nature Communications (online). Nov 6, 2015;**6**:7260

[73] Kim JY, Peck JH, Hwang SH, Hong J, Hong SC, Huh W, Lee SW. Preparation and mechanical properties of poly(vinyl chloride)/bamboo flour composites with a novel block copolymer as a coupling agent. Journal of Applied Polymer Science. 2008;**108**(4):2654-2659

[74] Hong B, Xue G, Weng L, Guo X. Pretreatment of moso bamboo with dilute phosphoric acid. BioResources. 2012;**7**(4):4902-4913

[75] Ladapo H, Oyegoke O, Bello R. Utilization of vast Nigeria's bamboo resources for economic growth: A review. Journal of Research in Forestry, Wildlife and Environment. 2017;**9**(2):29-35

[76] Naxium M. Biodiversity and resources exploitation of bamboo in China. In: Zhu Z ed. Sustainable Development of Bamboo and Rattan Sectors in Tropical China. INBAR and China Forestry Publishing House. Sector Proceedings. 2001. p. 6

[77] Xuhe C. Promotion of bamboo for poverty alleviation and economic development. Journal of Bamboo and Rattan. 2003;**2**(4):345-350

[78] Borah E, Pathak K, Deka B, Neog D, Borah K. Utilization aspects of bamboo and its market value. Indian Forester. 2008;**134**(3):423-427

[79] Choudhury D, Sahu JK, Sharma G. Value addition to bamboo shoots: A review. Journal of Food Science and Technology. 2012;**49**(4):407-414

[80] Ogunwusi A, Onwualu A. Indicative inventory of bamboo availability and utilization in Nigeria. Journal of Research in Industrial Development. 2011;**9**(2):1-9

[81] Parthiban M, Viju S. Bamboo charcoal yarns/fabrics – A new era in textile field. Textile Review. 2009;**3**:10-11

[82] Wang L-G, Yan G-B. Adsorptive removal of direct yellow 161dye from aqueous solution using bamboo charcoals activated with different chemicals. Desalination. 2011;**274**(1-3): 81-90

[83] Liu Y, Zhao X, Li J, Ma D, Han R. Characterization of bio-char from pyrolysis of wheat straw and its evaluation on methylene blue adsorption. Desalination and Water Treatment. 2012;**46**(1-3):115-123

[84] Yang Y-Y, Li Z-L, Wang G, Zhao X-P, Crowley DE, Zhao Y-H. Computational identification and analysis of the key biosorbent characteristics for the biosorption process of reactive black 5 onto fungal biomass. PLoS One. 2012;**7**(3):e33551

[85] Liao P, Ismael ZM, Zhang W, Yuan S, Tong M, Wang K, Bao J. Adsorption of dyes from aqueous solutions by microwave modified bamboo charcoal. Chemical Engineering Journal. 2012;**195**:339-346

[86] Lu K, Yang X, Shen J, Robinson B, Huang H, Liu D, Bolan N, Pei J, Wang H. Effect of bamboo and rice straw biochars on the bioavailability of Cd, Cu, Pb and Zn to Sedum plumbizincicola. Agriculture, Ecosystems & Environment. 2014;**191**:124-132

[87] Wang FY, Wang H, Ma JW. Adsorption of cadmium (II) ions from aqueous solution by a new low-cost adsorbent—Bamboo charcoal. Journal of Hazardous Materials. 2010;**177**(1-3):300-306

[88] Ma JW, Wang FY, Huang ZH, Wang H. Simultaneous removal of 2,4-dichlorophenol and Cd from soils by electrokinetic remediation combined with activated bamboo charcoal. Journal of Hazardous Materials. Apr 15, 2010;**176**(1):715-720

[89] Asada T, Ohkubo T, Kawata K, Oikawa K. Ammonia adsorption on bamboo charcoal with acid treatment. Journal of Health Science. 2006;**52**(5):585-589

[90] Liao P, Yuan S, Xie W, Zhang W, Tong M, Wang K. Adsorption of nitrogen-heterocyclic compounds on bamboo charcoal: Kinetics, thermodynamics, and microwave regeneration. Journal of Colloid and Interface Science. 2013;**390**(1):189-195

[91] Li Q, Pan H, Higgins D, Cao R, Zhang G, Lv H, Wu K, Cho J, Wu G. Metal–organic framework-derived bamboo-like nitrogen-doped Graphene tubes as an active matrix for hybrid oxygen-reduction electrocatalysts. Small. 2015;**11**(12):1443-1452

[92] Kuehl Y, Li Y, Henley G. Impacts of selective harvest on the carbon sequestration potential in Moso bamboo (*Phyllostachys pubescens*) plantations. Forests, Trees and Livelihoods. Jan 3, 2013;**22**(1):1-18

[93] Biopact Bamboo power. Indian state of Mizoram to produce electricity from bamboo. City. 2006. https://global.mongabay.com/news/bioenergy/2006/08/bamboo-power-indian-state-of-mizoram.html. [Accessed: 20 Sep, 2017]

[94] Macedo IC. The current situation and prospects for ethanol. Estudos Avançados. 2007; **21**(59):157-165

[95] Meneses TJB, Azzini A. A new feedstock for ethanol production. Boletim do ITAL. 1981; **18**:145-154

[96] Azzini A, de Arruda MCQ, Tomazello Filho M, Salgado ALB, Ciaramello D. Joint production of ethanol and cellulosic fibers from bamboo. Bragantia. 1987;**46**:17-25

[97] Zing-Ming L, Cheng L. Preparation and Morphology of Nano-Crystaline from Cellulose from Bamboo Pulp. Proceedings of 54th Society of Wood Science and Technology conference on sustainable development of wood and biomass in our new global economy, Beijing China: International Bamboo and Rattan. 2012:72

[98] Dharmananda S. Bamboo as Medicine. ITM; 2004. http://www.itmonline.org/arts/bamboo.htm [Accessed on 20th September 2017]

[99] Sharma P, Saikia P, Sarma K. Diversity, uses and in vitro propagation of different bamboos of Sonitpur District, Assam. Journal of Ecosystem & Ecography. 2016;**6**(2):1-9

[100] Majbauddin A, Kodani I, Ryoke K. The effect of bamboo leaf extract solution and sodium copper chlorophyllin solution on growth and volatile sulfur compounds production of oral malodor associated some anaerobic periodontal bacteria. Yonago Acta Medica. 2015;**58**(3):129

[101] Akinbile C, Fakayode O, Sanusi K. Using bamboo (*Bambusa vulgaris*) as a field drainage material in Nigeria. African Journal of Environmental Science and Technology. 2011; **5**(12):1124-1127

[102] Lipangile T. Manufacture and construction of bamboo water supply systems. Journal of American Bamboo Society. 1991;**8**:191-198

[103] McClure FA. The Bamboos. London: Harvard University Press; 1967

[104] Ogunwusi A. Potentials of bamboo in Nigeria's industrial sector. Journal of Research in Industrial Development. 2011;**9**(2):136-146

[105] Ogunwusi A. Forest products industry in Nigeria. African Research Review. 2012;**6**(4): 191-205

[106] Azzini A, Gondim-Tomaz RMA. Extração de amido em cavacos de bambu tratados com solução diluída de hidróxido de sódio Starch extraction from bamboo chips treated with sodium hydroxide diluted solution. Bragantia. 1996;**55**(2):215-219

[107] Miura M, Watanabe I, Shimotori Y, Aoyama M, Kojima Y, Kato Y. Microbial conversion of bamboo hemicellulose hydrolysate to xylitol. Wood Science and Technology. 2013;**47**(3):515-522

[108] Parajó J, Garrote G, Cruz J, Dominguez H. Production of xylooligosaccharides by autohydrolysis of lignocellulosic materials. Trends in Food Science & Technology. 2004;**15**(3-4):115-120

[109] Nabarlatz D, Ebringerová A, Montané D. Autohydrolysis of agricultural by-products for the production of xylo-oligosaccharides. Carbohydrate Polymers. 2007;**69**(1):20-28

[110] Vazquez M, Alonso J, Domınguez H, Parajo J. Xylooligosaccharides: Manufacture and applications. Trends in Food Science & Technology. 2000;**11**(11):387-393

[111] Aoyama M, Seki K. Chemical characterization of solubilized xylan from steamed bamboo grass. European Journal of Wood and Wood Products. 1994;**52**(6):388-388

[112] Aoyama M, Seki K. Acid catalysed steaming for solubilization of bamboo grass xylan. Bioresource Technology. 1999;**69**(1):91-94

[113] Chongtham N, Bisht MS, Haorongbam S. Nutritional properties of bamboo shoots: Potential and prospects for utilization as a health food. Comprehensive Reviews in Food Science and Food Safety. 2011;**10**(3):153-169

[114] Waite M, Platts J. Engineering sustainable textiles: A bamboo textile comparison. Energy, Environment, Ecosystems, Development and Landscape Architecture. 2009;**3**(5):362-368

[115] Mekuriaw Y, Urge M, Animut G. Role of indigenous bamboo species (*Yushania alpina* and *Oxytenanthera abyssinica*) as ruminant feed in northwestern Ethiopia. Livestock Research for Rural Development. 2011;**23**(9):9

[116] Hu C, Zhang Y, Kitts DD. Evaluation of antioxidant and prooxidant activities of bamboo Phyllostachys nigra var. Henonis leaf extract in vitro. Journal of Agricultural and Food Chemistry. 2000;**48**(8):3170-3176

[117] Jiao J, Zhang Y, Liu C, Liu J e, Wu X, Zhang Y. Separation and purification of tricin from an antioxidant product derived from bamboo leaves. Journal of Agricultural and Food Chemistry. 2007;**55**(25):10086-10092

[118] Xie J, Lin Y-S, Shi X-J, Zhu X-Y, Su W-K, Wang P. Mechanochemical-assisted extraction of flavonoids from bamboo (*Phyllostachys edulis*) leaves. Industrial Crops and Products. 2013;**43**:276-282

[119] Suzuki T, Jacalne DV. Above-ground biomass and the growth of bamboo stands in the Philippines. JARQ. 1986;**20**(1):85-91

Chapter 4

Preliminary Safety Evaluation of Bamboo Pyrolysis Products: Charcoal and Vinegar

Han Chien Lin

Additional information is available at the end of the chapter

http://dx.doi.org/10.5772/intechopen.68542

Abstract

The bamboo charcoal is manufactured in earth kilns with the temperature at 700–800°C from Moso bamboo (*Phyllostachys heterocycla*). Bamboo vinegars, by-products of the charcoal, are collected from the chimney outflow of earthen kiln at six different temperatures at 80–150°C, and with categories over 80, 90–92, 99–102, 120–123, and 145–150°C during the pyrolysis of the charcoal. The preliminary safety evaluation using the Ames test for the bamboo charcoals has no cytotoxicity and mutagenicity toward *Salmonella typhimurium* TA98 and TA100 with S9 mix and without S9 mix. This suggests that the charcoal can not only be considered to be a safe pigment for food but also be used as a natural moisture-proof material. The safety of the bamboo vinegars shows that neither cytotoxicity nor mutagenicity toward *S. typhimurium* TA98 and TA100 with S9 mix at the diluting percent content of vinegars is lower than 20.00% or less and without S9 mix is at 33.33% or less, and the reverse mutation assay (antimutagenic activity) denotes that the vinegars express this dose-dependent inhibitory effect against both 4-nitroquinoline-*N*-oxide and aflatoxin B_1 in *S. typhimurium* TA98 and TA100. The main compounds in the vinegars may partially account for the safety evaluation of biological action.

Keywords: pyrolysis, bamboo charcoal, bamboo vinegar, Ames test, mutagenicity

1. Introduction

Bamboo is regarded as a natural and renewable bioresource, a green and environmentally friendly material, and has chemical and physical qualities/properties similar to wood. It quickly reaches its full potential size, which usually only takes approximately 60–70 days after the bamboo culms emerge from the ground [1]. According to the annual statistics of the agricultural reports in Taiwan (2015), the volume of diverse bamboos is about 1,581,330,000 pieces. The three main types of bamboo used include *Phyllostachys makinoi* Hayata (Makino bamboo), *Phyllostachys heterocycla* Milf (Moso bamboo), and *Dendrocalamus latiflorus* Munro (Ma bamboo) [2]. This indicates that bamboo is one of the economic green resources in Taiwan. Among these bamboos, Moso bamboo has been mainly used as a material for the making of furniture, handicrafts, and athletic/leisure goods, because it can grow over 20 m tall and 60–150 mm in diameter in one growing season [3]. Moreover, bamboo is accumulated from organic compounds from which a bamboo converts carbon dioxide (CO_2) by photosynthesis, but it can naturally decay by organic compound oxidization or be burned, and this produces CO_2 that returns to the atmosphere. To decrease bamboo decaying or burning, the preparation of charcoal becomes one of the selected methods because carbon can extend the lifetime of bamboo [4]. In other words, the charcoals can also become earth friendly materials because they can slow down the increase of CO_2 concentration in the atmosphere.

Bamboo and its products show good prospects for commercial applications, when considering the need for the protection of our wood resources and environmental balance. It is therefore important to study the characteristics of bamboo and its by-products in order to make good use of them. Domestic and foreign manufacturers and researchers have invested a great deal of money, labor and time to discover the characteristics and functions of charcoal in recent years. Bamboo charcoals are mainly derived from 4-year-old or older bamboo as raw materials [4–7]. The utilization of bamboo charcoal, conventionally regarded as fuel, is widely applied to daily life or/and industry. The use of bamboo charcoal is more wide and diverse because it is a porous material with a high specific surface area that has lots of functions, such as indoor deodorization, humidity control, water quality improvement, air purification and so on [5, 8, 9]. Recently, there has been a tendency to maintain good health from food products. Some food producers have added charcoal materials into food products, for example, charcoal bread/ cookies, charcoal peanuts, charcoal ice cream, etc. This is advertised as being able to absorb unclean substances, such as heavy metal elements, and producers have exaggerated that these materials can clean the intestines and stomach after eating. However, in May of 2006, the Department of Health's Executive Yuan, Taiwan, announces that the charcoals can only be used as colorants of food without any medical and health effects, that is, as a natural black pigment only [10]. The charcoals can be added as a pigment in food, but it is a profound question whether or not residue *in vivo* causes any harm by inducing cell lesions or carcinogens.

The application of bamboo vinegar (brown-red transparent liquids), even the compounds that are complex and different, is mainly able to be divided into three main portions: acid, phenol, and neutral compounds [11]. The vinegar consists of 80–200 compounds: 32% organic acid, 40% phenolic compound, 3% aldehyde, 5% alkone compound, 5% alcohol compound, 4% ester compound, and 5% others. When bamboo vinegar is dehydrated, there is usually 80% water [12–15]. The organic compounds in bamboo vinegars may have practical applications even when present in only trace quantities [14, 16], such as in improving soil, promoting crops and preventing worm growth, as well as reducing agricultural chemicals, compost odor and sterilization [12]. Recently, bamboo vinegars have been developed that are beneficial for promoting growth of plants to as a plant root growth promoter or a pH value adjuster of cultural media [17–19]. It is also effective when used against allergies [20], in healthy drinks [13, 21, 22], as a virus/fungi/bacterial resistant [16, 23–27] and as an agent of antioxidation, especially for a resistant lipid oxidation effect [15]. As stated in the above references, the commercial production of bamboo vinegar is being increased and highly valued for its diverse effective uses in Taiwan. However, bamboo vinegars collect at the exit of chimney of earthen or furnace kiln, when the carbonization temperature of bamboo is raised to over 500°C, produce some carcinogenic polycyclic aromatic hydrocarbons (PAHs), such as naphthalene and phenanthrene [28–30]. The concentrations of these toxics increase with the increase in temperature as well. Even though the council of agriculture in Taiwan has submitted certified agricultural standards of forest products (2004) to prove that it is necessary for bamboo vinegars to be collected at 80–150°C at an exit of chimney of earthen kiln and below 350°C for a furnace kiln [31], the collected bamboo vinegars from above this range of temperatures are necessary to evaluate the potential of mutagenic and carcinogenic agents, due to the fact that they are omnipresent in the human environment and seem impossible to completely eliminate.

Ames *et al.* [32] reports that for screening of environmental mutagens and carcinogens, the Ames test (safety evaluation), a convenient method to evaluate mutagenic activities of these chemicals, has been developed, and McCann *et al.* and Sugimura *et al.* have suggested that the mutagenic activities of a number of chemicals correlate well with the carcinogenic activities [33, 34]. The main goal of this chapter is to realize fundamental knowledge of the bamboo charcoal/vinegars as functional additives, while at the same time understanding the preliminary safety evaluation of both using the bacterial mutation assay on *Salmonella typhimurium* (*S. typhimurium*) TA98 and TA100 strains with and without an extrinsic metabolic activation system. Moreover, because the reverse mutation assay (antimutagenic activity) has an array of prospective applications in human care, such as the increasing application in drinks and food antioxidation, and has not been reported for the antimutagenic activities of bamboo vinegars that have been made so far, the antimutagenic activity of the vinegars is reported. Besides, the basic compounds of vinegars, collected at different temperatures from the exit of earthen kiln chimney and analyzed by using gas chromatography-mass spectroscopy, are considered, in order to realized the effect of bamboo vinegars' compounds on preliminary safety evaluation.

2. Ames test for bamboo charcoal and vinegars

The preliminary safety evaluation, including cytotoxicity and mutagenicity, is performed in accordance with the Ames test [32], a widely used convenient and short-term assay with predictable accuracy for carcinogen up to 72–91% [35]. Referring to the Ames test, as proposed by Ames *et al.* [32], Waleh *et al.* [36], which indicates that if the sample is toxic for the strain, the bacterial count decreases, and the test result is likely to be misjudged. Therefore, the specimen cytotoxicity has to be tested before the mutagenicity test, in order to evaluate whether the growth of the strain is influenced. The mutagenicity can be implemented if there is no toxicity [32, 36]. In this report, the *S. typhimurium* TA98 and TA100 are used as test strains. TA98 is the strain sensitive to frame shifting mutation, as caused by specific mutagens; TA100 is the strain for testing base substitution variation [37].

2.1. Cytotoxicity

The methods of the cytotoxicity for the bamboo charcoal and vinegars are taken 1.0, 2.5, 5.0, 7.5, and 10.0 mg of bamboo charcoal, as well as 0.1 mL of bamboo vinegars, collected at different temperatures (80–150, over 80, 90–92, 99–102, 120–123, and 145–150°C), from the exit of chimney of earthen kiln are diluted to a percent content of 50, 33.33, 25, 20, 13.33, and 10%, respectively. Both of them are examined with *S. typhimurium* TA98 and TA100 for either S9 (+S9) or zero S9 (-S9) in accordance with the Ames test and the experimental procedure referred to [32, 37–39]. The colony count is calculated; if the bacterial count of the test group (+S9 or -S9) is larger than the bacterial count of the control group (no bamboo charcoal/vinegars) by 80% (the bacterial count rate, Survival), there is no toxicity [35]. The Survival (%) is the residual bacteria rate that is the percentage relative to the control (100%). The formula of the Survival of cytotoxicity is: Survival (%) = (the bacterial count of test group/the bacterial count of control group) × 100.

2.2. Mutagenicity

The mutagenicity is analyzed by using the method proposed by Maron and Ames [40]. The test charcoal and vinegars for this mutagenicity test, with or without S9 mix, are the same as for the cytotoxicity test, and the experimental procedure is referred by [37–39]. If the colony count of the TA98 and TA100 test group is larger than the control group by more than two times; that is, the mutagenicity ratio is larger than 2, the specimen for bamboo charcoal/vinegars is considered to have mutagenicity. The mutagenicity ratio is calculated as: mutagenicity ratio (MR) = induced revertants per plate/spontaneous revertants per plate (blank).

2.3. Antimutagenic activity

The test vinegars of the antimutagenic activity are assayed according to the Ames method [40]. The mutagens are 4-nitroquinoline-*N*-oxide (NQNO) and aflatoxin B_1 (AFB_1) that are diluted with dimethyl sulfoxide (DMSO). NQNO (1 μg/plate for TA98 and TA100, respectively), a direct-acting mutagen and AFB_1 (5 μg/plate for TA98 and TA100, respectively), an indirect mutagen which requires metabolic activation, require S9 mix for metabolic activation, respectively. A mutagen (0.1 mL; contained 1 μg NQNO or 5 μg AFB_1) is added to the mixture of a strain (TA98 or TA100), and 0.1 mL of each test vinegar is added to S9 mix for AFB_1 or to the phosphate buffer (0.1 mol/L, pH 7.4) for NQNO. The mutagenicity of each mutagen in the absence of an extract is defined as 100%. The inhibition (%) of mutagenicity of test vinegar is calculated as follows:

Inhibition (%) = [1–(number of His^+ revertants in the presence of the test vinegar—number of spontaneous revertants)/(number of His^+ revertants in the absence of the test vinegar—number of spontaneous revertants)] × 100.

3. Safety evaluation of bamboo charcoal

3.1. Basic properties of bamboo charcoal

The true density, BET specific surface area and average pore diameter of Moso bamboo charcoals are 1.68 (g/cm^3), 138.70 (m^2/g), and 2.41 (nm), respectively. Moreover, the heavy metal element of bamboo charcoal for Br, Pb, Hg, Cr and Cd is analyzed using X-ray Fluorescent Analyzer of XGT-1000WR [39]. Br and Cr are not detected because their amounts are probably very low. Both Cd and Hg in the charcoal are closed to 0.5 ppm. The Pb in the charcoal is 2.9 ppm. That meets the Sanitation Standard for Edible Natural Colorants, Food Sanitation Standards (1989), Ministry of Health and Welfare in Taiwan at below 40 ppm [41].

3.2. Cytotoxicity of bamboo charcoal

The cytotoxicity test results with 1.0, 2.5, 5.0, 7.5 and 10.0 mg of Moso bamboo charcoal for *S. typhimurium* TA98 and TA100 are shown in **Figure 1**. The Survival (%) of the Moso bamboo charcoal with either zero S9 (-S9) or S9 (+S9) is higher than 80%. Waleh *et al.* [36] indicate that the amount of residual bacteria of S. *typhimurium* must be over 80% of the control group (Control) to determine that the test group has no cytotoxicity for S. *typhimurium* [36]. The Survival of the charcoals is higher than that of control by more than 80%, indicating that the Moso bamboo charcoal has no cytotoxicity for the test strains in the additional range of 1–10 mg/plate, and the dose for the mutagenicity test can be selected according to this range.

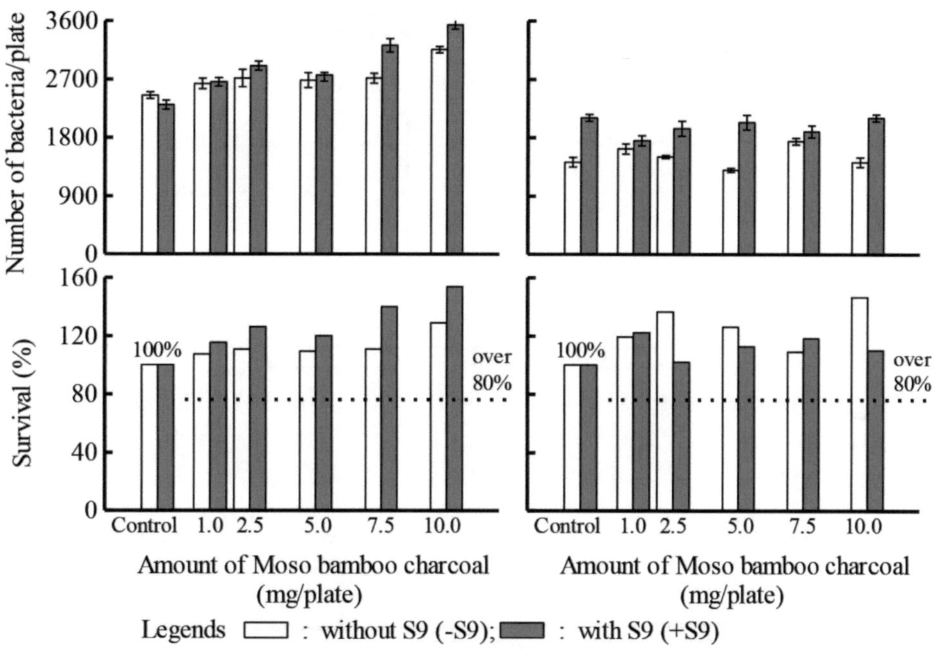

Figure 1. Cytotoxicity of Moso bamboo charcoal toward *S. tyhpimurium* TA98 and TA100 without or with S9 mix.

3.3. Mutagenicity of bamboo charcoal

Figure 2 shows the mutagenicity test results of the Moso bamboo charcoal for *S. typhimurium* TA98 and TA100. The bamboo charcoal, without or with S9, in the test range (1–10 mg/plate) does not exceed spontaneous revertants by more than two times for TA98 and TA100, that is, the mutagenicity ratio (MR) is smaller than 2. According to the standards proposed by Ames *et al.* [32], if the number of spontaneous revertants induced by the specimen is larger than the spontaneous revertants of the control group by more than two times, the specimen has mutagenicity. Therefore, the bamboo charcoals have no mutagenicity toward *S. typhimurium* TA98 and TA100. Furthermore, Weng reports that the maximum percent weight of Moso bamboo at RH 40% is 6.7% and at RH 90% is 9.9% [39]. The water activity of the bamboo charcoal (0.57), compared to Aw for the growth of a microorganism environment, is below 6.0. The heavy metal contents (as Pb ppm base) of charcoal are below 40 ppm and meet the Sanitation Standard for Edible Natural Colorants, Food Sanitation Standards as well. For the Ames test, the bamboo charcoal has no toxicity and mutagenicity toward *S. typhimurium* TA98 and TA100. In the scope of this report, the above results indicate that the bamboo charcoal can be expected to be the same as the materials of edible natural colorants.

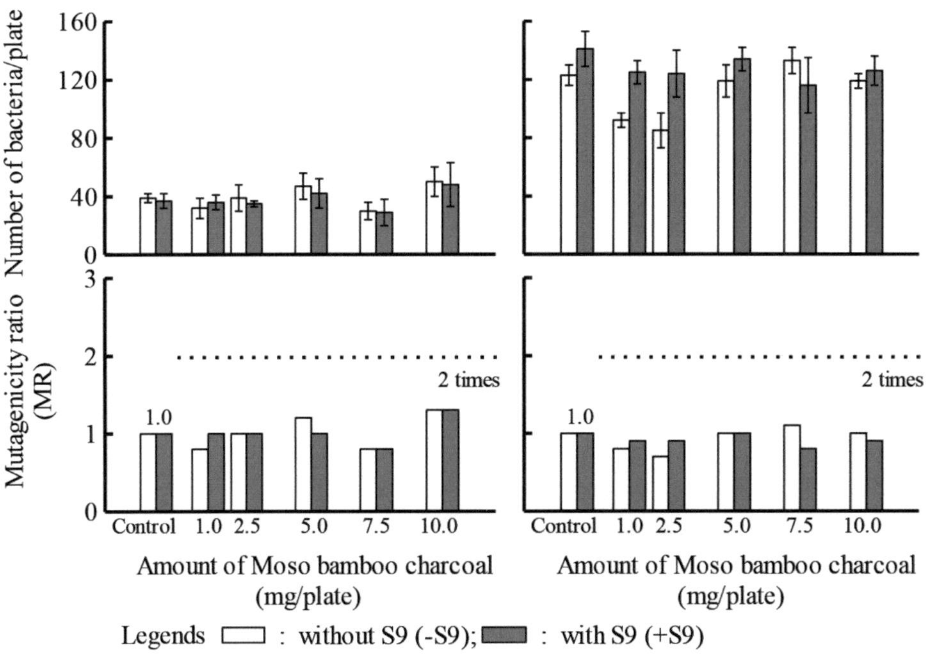

Figure 2. Mutagenicity of Moso bamboo charcoal toward *S. tyhpimurium* TA98 and TA100 without or with S9 mix.

4. Safety evaluation of bamboo vinegars

4.1. Cytotoxicity of bamboo vinegars

One of the cytotoxicity results for the bamboo vinegars collected at temperatures of 90–92°C with the original vinegar (no diluting) and a range of diluting percent contents of 50.00, 33.33, 25.00, 20.00, 13.33 and 10.00% for *S. typhimurium* TA98 and TA100 without S9 mix, as well as for with S9 mix with diluting percent contents of 33.33, 25.00, 20.00, 13.33 and 10.00% is shown in **Table 1**.

The residual bacterial count of the control group (Blank) toward *S. typhimurium* TA98 and TA100 is 1838 and 1500, respectively. For bamboo vinegars collected at temperatures of

S9 mixture	Diluting percent content (%)	Bamboo vinegar (90–92°C)			
		TA98	Survival[1] (%)	TA100	Survival (%)
	Blank[2]	**1838[3]**	**100.00**	**1500**	**100.00**
–S9	Original vinegar	0	0.00	87	5.82
	50.00	336	18.26	285	19.00
	33.33	1652	89.86	1284	85.58
	25.00	1822	99.13	1201	80.04
	20.00	1736	94.43	1264	84.27
	13.33	1829	99.53	1353	90.20
	10.00	1830	99.56	1641	109.42
S9	Blank	2165	100.00	2585	100.00
	33.33	1605	74.16	1736	67.15
	25.00	1960	90.53	2209	85.46
	20.00	1938	89.53	2243	86.75
	13.33	2046	94.53	2885	111.60
	10.00	2000	92.39	2249	87.00

[1]Survival (%) = (the bacterial count of test group/the bacterial count of control group) × 100.
[2]Blank (the control group) was added without bamboo vinegars.
[3]Mean.

Table 1. Cytotoxicity of bamboo vinegar collected at the temperature of 90–92oC toward *Salmonella typhimurium* TA98 and TA100 without or with S9 mix.

90–92°C, the residual bacterial count without S9 for diverse diluting percent contents is 0–1830 for TA98 and 87–1641 for TA100; for those with S9, it is 1605–2000 for TA98 and 1736–2249 for TA100. Waleh *et al.* indicate that the residual bacteria rate of *S. typhimurium* must be over 80% of the control group to determine that the test group has no cytotoxicity for *S. typhimurium* [36]. The residual bacteria rate (Survival, %) toward TA98 and TA100 for bamboo vinegar without S9 mix at a diluting percent content 33.3% or less, and with S9 mix at 25.00% or less is greater than 80%. In other words, the bamboo vinegars collected at temperatures of 90–92°C has cytotoxicity for *S. typhimurium* at a diluting 33.3% less. For vinegars collected at all different temperatures, the Survivals are shown in **Table 2**.

The Survival of all bamboo vinegars collected from 80 to 150, over 80, 90–92, 99–102, 120–123, and 145–150°C without S9 mix at a diluting percent content of 33.33% or less is all higher than those for Blank by more than 80%. However, the Survival of the bamboo vinegar collected at temperatures of 99–102°C at a diluting percent content of 25% shows that the cytotoxicity toward *S. typhimurium* TA98 and TA100 with S9 mix is lower than 80%, indicating "with toxicity." It is asserted that the Survival of bamboo vinegars collected at all different temperatures at a diluting percent content of 20.00% less for either with or without S9 mix is no cytotoxicity absolutely, and the dose for the mutagenicity test can be selected according to the results (**Table 2**).

Specimens	Diluting percent content (%)		Collection temperature (°C)					
			Over 80		90–92		99–102	
			TA98	TA100	TA98	TA100	TA98	TA100
Bamboo vinegars	−S9	33.33	95.16[1]	85.42	89.86	85.58	91.69	90.18
		25.00	115.54	88.91	99.13	80.04	99.13	105.82
		20.00	86.43	109.11	94.43	84.27	84.64	86.20
		13.33	95.30	88.11	99.53	90.20	99.53	90.56
		10.00	106.58	94.69	99.56	109.42	101.16	84.53
	S9	25.00	88.81	92.83	90.53	85.46	76.36	70.81
		20.00	95.40	84.84	89.53	86.75	86.53	106.65
		13.33	92.12	120.27	94.53	111.60	98.66	104.95
		10.00	84.19	99.79	92.39	87.00	94.73	83.29
Specimens	**Diluting percent content (%)**		**120–123**		**145–150**		**80–150**	
			TA98	**TA100**	**TA98**	**TA100**	**TA98**	**TA100**
Bamboo vinegars	−S9	33.33	109.12	90.76	95.92	112.16	109.79	104.98
		25.00	104.13	115.42	90.86	99.82	112.82	102.42
		20.00	111.19	115.69	107.04	123.40	124.23	108.62
		13.33	114.93	118.91	109.61	102.42	95.88	128.20
		10.00	113.75	111.47	122.63	96.44	127.98	119.71
	S9	25.00	89.79	96.39	103.87	122.07	86.94	89.12
		20.00	113.32	90.36	89.11	89.27	85.19	89.74
		13.33	126.73	115.78	143.39	140.28	84.85	97.83
		10.00	117.83	99.02	120.54	98.09	102.71	131.10

Unit: %.

[1]Survival (%) = (the bacterial count of test group/the bacterial count of control group) × 100.

Table 2. Survival of diverse bamboo vinegars collected at different temperatures toward *Salmonella typhimurium* TA98 and TA100 without or with S9 mix.

4.2. Mutagenicity of bamboo vinegars

Ames *et al.* report that the number of spontaneous revertants induced by the specimen is less than that for the control group by more than two times; the specimen has no mutagenicity [32]. **Table 3** shows the mutagenicity results for the bamboo vinegar collected at temperatures of 90–92°C with a diluting percent content of 25.00, 20.00, 13.33 and 10.00% for *S. typhimurium* TA98 and TA100 without S9 mix and with S9 for a diluting percent content of 20.00, 13.33 and 10.00%. The spontaneous revertants of Blank are 65 for TA98 and 132 for TA100 and of bamboo vinegars without S9 mix are from 41 to 53 for TA98 and from 109 to 129 for TA100. With S9 mix, the spontaneous revertants of Blank and bamboo vinegars are 69–76 for TA98 and 140–206 for TA100. The results on the same **Table 3** also show that the bamboo vinegars in the test range do not exceed spontaneous revertants by more than two times for TA98 and TA100 without/with S9 mix; that is, the MR is less than 2. The spontaneous revertants of the other bamboo vinegars collect at diverse temperatures, 80–150, over 80, 99–102, 120–123 and 145–150°C, at the ranges of the diluting percent content are smaller than those of Blank by more than two times [38].

S9 mixture	Diluting percent content (%)	Bamboo vinegar (90–92°C)			
		TA98	MR[1]	TA100	MR
	Blank[2]	65[3]	1.00	132	1.00
–S9	25.00	53	0.81	129	0.98
	20.00	42	0.64	121	0.91
	13.33	53	0.81	109	0.83
	10.00	41	0.63	115	0.87
S9	Blank	76	1.00	140	1.00
	20.00	69	0.90	195	1.39
	13.33	73	0.96	204	1.45
	10.00	74	0.97	206	1.47

[1]MR (mutagenicity ratio) = induced revertants per plate/spontaneous revertants per plate (control).
[2]Blank (the control group) was added without either bamboo or wood vinegars.
[3]Mean.

Table 3. Mutagenicity of diverse bamboo vinegar collected at the temperature of 90–92°C toward *Salmonella typhimurium* TA98 and TA100 without or with S9 mix.

The MR of the bamboo vinegars collected at all different temperatures toward *S. typhimurium* TA98 and TA100 without S9 mix for a diluting percent content of 25.00, 20.00, 13.33 and 10.00% or with S9 mix for a diluting percent content of 20.00, 13.33 and 10.00% is shown in **Table 4**. No matter what the vinegars are for all collected temperatures and *S. typhimurium* TA98 and TA100 without/with S9 mix, the MR is less than 2. Neither mutagenicity nor toxicity is observed for bamboo vinegars collected at diverse temperatures toward *S. typhimurium* TA98 or TA100 without/with S9 mix. The bamboo vinegars, therefore, have no mutagenicity toward *S. typhimurium* TA98 at 25% less and TA100 at 20% less.

4.3. Antimutagenic activity of bamboo vinegars

If mutagenicity occurred in a treated material, the results of the antimutagenic assay would be confused due to increased or decreased numbers of revertants of TA98 and TA100 [42]. Hence, without S9 mix a diluting percent content of 25.00–10.00% and with S9 mix at 20.00–10.00% a diluting percent content are selected for the antimutagenic activity. NQNO and AFB_1 are used as direct mutagens requiring metabolic activation and indirect acting mutagen, respectively

Specimens	Diluting percent content (%)		Collection temperature (°C)					
			Over 80		90–92		99–102	
			TA98	TA100	TA98	TA100	TA98	TA100
Bamboo vinegars	–S9	25.00	0.72[1]	1.08	0.81	0.98	–	–
		20.00	0.68	1.02	0.64	0.91	0.65	0.90
		13.33	0.70	0.86	0.81	0.83	0.72	0.78
		10.00	0.61	0.93	0.63	0.87	0.75	1.00
	S9	20.00	1.10	1.47	0.90	1.39	0.88	1.57
		13.33	0.91	1.51	0.96	1.45	0.89	1.62
		10.00	1.04	1.42	0.97	1.47	0.96	1.66
Specimens	**Diluting percent content (%)**		**120–123**		**145–150**		**80–150**	
			TA98	**TA100**	**TA98**	**TA100**	**TA98**	**TA100**
Bamboo vinegars	–S9	25.00	0.74	0.63	0.68	0.50	0.62	0.80
		20.00	0.81	0.62	0.59	0.54	0.67	0.76
		13.33	0.52	0.57	0.57	0.50	0.68	0.72
		10.00	0.69	0.77	0.69	0.56	0.73	0.60
	S9	20.00	1.18	1.19	1.32	1.32	1.16	0.83
		13.33	1.14	1.29	1.22	1.26	1.21	1.04
		10.00	1.00	1.28	1.20	1.40	1.35	1.05

Unit: %.
[1]MR (mutagenicity ratio) = induced revertants per plate/spontaneous revertants per plate (control).

Table 4. Mutagenicity ratio of diverse bamboo vinegars collected at different temperatures toward *Salmonella typhimurium* TA98 and TA100 without or with S9 mix.

[37]. Doses of mutagens, 1 μg for NQNO and 5 μg for AFB_1, are selected from a dose-response curve of a preliminary experiment [43]. The His^+revertants of strain are less than those of the control group (Blank), indicating that with antimutagenic activities. Meanwhile, the inhibitory effect of the specimen is expressed by inhibition (%), and the higher the inhibition, the more effective the antimutagenic activities [40]. The inhibitory effects for one of the antimutagenic activity results for the bamboo vinegar collected at temperatures of 90–92°C with a diluting percent content of 25.00, 20.00, 13.33 and 10.00% for NQNO and at 20.00, 13.33 and 10.00% for AFB_1 are summarized in **Table 5**.

The His^+ revertants of strain against the NQNO in Blank (without bamboo vinegars) are 1128 for TA98, and 1445 for TA100, for AFB_1: they are 1824 for TA98 and 2406 for TA100. The spontaneous revertants without NQNO are 65 for TA98 and 132 for TA100 and without AFB_1 are 76 for TA98 and 140 for TA100. The His^+ revertants of strain (inhibition,%) against the NQNO for bamboo vinegar at 90–92°C with different diluting percent contents are 807–985 (30.18–13.49%) for TA98 and 425–616 (77.65–63.15%) for TA100. For AFB_1, they are 1114–1319 (40.63–28.88%)

Mutagens	Diluting percent content (%)	Bamboo vinegar (90–92°C)			
		TA98	Inhibition[1] (%)	TA100	Inhibition (%)
NQNO (1 μg/plate)	Blank[2]	1128[3]	0.00	1445	0.00
	25.00	807	30.18	425	77.65
	20.00	841	27.04	446	76.05
	13.33	891	22.33	594	64.80
	10.00	985	13.49	616	63.15
Spontaneous revertants		65		132	
AFB_1 (5 μg/plate)	Blank	1824	0.00	2406	0.00
	20.00	1114	40.63	843	68.97
	13.33	1292	30.44	1012	61.53
	10.00	1319	28.88	1202	53.14
Spontaneous revertants		76		140	

[1]Inhibition (%) = [1–(number of His^+revertants in the presence of the test vinegar—number of spontaneous revertants)/(number of His^+revertants in the absence of the test vinegar—number of spontaneous revertants)] × 100.

[2]Blank (the control group) is added without either bamboo or wood vinegars.

[3]Mean.

Table 5. Antimutagenic activity of diverse bamboo vinegar collected at the temperature of 90–92°C toward *Salmonella typhimurium* TA98 and TA100 with S9 mix.

for TA98 and 843–1202 (68.97–53.14%) for TA100 from bamboo vinegar at 90–92°C with different diluting percents. The results also show that the higher diluting percent content, the greater the inhibition, as well as, no matter what the vinegar is, the inhibition for TA100 is greater than that of TA98. Moreover, the inhibition of the bamboo vinegars collected at all different temperatures against the NQNO for diluting percent contents of 25.00, 20.00, 13.33 and 10.00% or against the AFB_1 at 20.00, 13.33 and 10.00% is shown in **Table 6**.

The inhibition of the bamboo vinegars to TA98 is 0.19–30.18% for NQNO and 0.92–40.63% for AFB_1. For TA100 against NQNO (18.18–81.77%) and AFB_1 (37.43–75.24%), they are better than those for TA98. The antimutagenic activity to NQNO is effective for bamboo vinegars collected at diverse temperatures with a diluting percent content of 25.00% or less, and for AFB_1, it is also effective at a 20.00% or less diluting percent content. Furthermore, the bamboo vinegars show that the inhibitory effect on NQNO or AFB_1 toward TA100 is greater than that toward TA98. It is also indicated that the inhibition of the vinegars against AFB_1 toward TA98 and TA100 is better than that against NQNO.

4.4. Effect of bamboo vinegars' compounds on preliminary safety evaluation

The identified compounds of bamboo vinegars collected from different temperatures in the category of 80–150°C, over 80, 90–92, 99–102, 120–123 and 145–150°C are analyzed by gas chromatography-mass spectroscopy [15, 38, 44]. The acid, phenol, ketone and other

Specimens	Diluting percent content (%)		Collection temperature (°C)					
			Over 80		90–92		99–102	
			TA98	TA100	TA98	TA100	TA98	TA100
Bamboo vinegars	NQNO	25.00	25.03[1]	70.14	30.18	77.65	-	-
		20.00	23.15	65.34	27.04	76.05	23.21	75.22
		13.33	12.80	60.64	22.33	64.80	20.58	58.91
		10.00	11.29	59.45	13.49	63.15	15.18	57.97
	AFB_1	20.00	24.38	62.94	40.63	68.97	37.80	66.35
		13.33	21.13	57.88	30.44	61.53	26.36	66.00
		10.00	18.31	37.43	28.88	53.14	24.26	54.88
Specimens	**Diluting percent content (%)**		**120–123**		**145–150**		**80–150**	
			TA98	**TA100**	**TA98**	**TA100**	**TA98**	**TA100**
Bamboo vinegars	NQNO	25.00	27.85	74.45	2.76	81.77	22.96	68.97
		20.00	11.67	64.45	1.19	60.23	4.20	61.38
		13.33	4.08	61.55	1.00	40.71	0.75	57.31
		10.00	1.07	28.52	0.63	18.18	0.19	42.74
	AFB_1	20.00	20.03	75.24	11.63	73.06	20.14	74.86
		13.33	18.73	68.97	8.54	66.41	17.09	73.27
		10.00	13.20	59.88	0.92	60.50	7.44	58.44

[1]Inhibition (%) = [1–(number of His^+revertants in the presence of the test vinegar—number of spontaneous revertants)/(number of His^+revertants in the absence of the test vinegar—number of spontaneous revertants)] × 100.

Table 6. Inhibition of diverse bamboo vinegars collected at different temperatures toward *Salmonella typhimurium* TA98 and TA100 with S9 mix.

compounds of bamboo vinegars are about 10.65–20.09%, 57.87–65.98%, 10.13–18.76% and 9.66–15.30, respectively. The acid compounds included butanoic acid, 2-methoxyethyl acetate, 4-hydroxy-butanoic acid and 4-hydroxy-3-methoxy-butanoic acid. The maximum fraction of acid compounds is the bamboo vinegar collected from 80 to 150°C. The phenol (5.93–16.60%), 2-methoxy-phenol (8.27–16.39%) and 4-ethyl-phenol (3.68–9.48%) are the main fractions of phenol compounds for bamboo vinegars. For ketone compounds, the 2-hydroxy-3-methyl-2-cyclopentenone-1-one, 2, 3-dimethyl-2-cyclopentenone-1-one and maltol can be measured for bamboo vinegars collected at all temperatures. The maximum fraction of ketone compounds is the bamboo vinegar collected from 120 to 123°C.

According to the former, results of safety evaluation (**Tables 1** and **2**; **Tables 3** and **4**) and antimutagenic activity (**Tables 5** and **6**) are present in diverse bamboo vinegars, the mutagenicity is occurred in the diluting percent content of vinegars that are higher than 20.00%, and without S9 mix is at 33.33%, but the diluting percent content of vinegars of 20.00% or less, expressed the amount-dependent inhibitory effect against both NQNO with 1 μg/plate and AFB_1 with 5 μg/plate in *S. typhimurium* TA98 and TA100. This is because the antimicrobial activity is affected by some phenol compounds [45]. In this report, the main fractions of bamboo vinegars are phenol compounds, 57.87–65.98% [38]. It is inferred that the phenol compounds may partially account for the biological action of bamboo vinegars.

5. Conclusion

The biological action of bamboo charcoal and vinegars is evaluated by *Salmonella* mutagenesis assay, as a safety evaluation and antimutagenic activity. The charcoal can be expected to be similar to the materials of edible natural colorants and/or natural drying agents, because the water activity of charcoal is below 6.0, the Pb base of charcoal is below 40 ppm, and the bamboo charcoals have no toxicity and mutagenicity toward *Salmonella typhimurium* TA98 and TA100. For the bamboo vinegars, the main percent content of the phenol compounds for bamboo vinegars is phenol (5.93–16.60%) and 2-methoxy-phenol (8.27–16.39%). The diluting percent content of vinegars is lower than 20.00% or less with S9 mix and 33.33% or less without S9 mix in cytotoxicity and mutagenicity toward *Salmonella typhimurium* TA98 and TA100 because the rest of the bacterium (Survival) these percent contents is higher than 80% of the control group (Blank), and the mutagenicity ratio (MR) is less than for the control group by more than two times. The diluting percent content of vinegars of 20.00% or less expressed the amount-dependent inhibitory effect against both the mutagenicity of 4-nitroquinoline-N-oxide (NQNO) with 1 μg/plate and aflatoxin B1 (AFB_1) with 5 μg/plate in *Salmonella typhimurium* TA98 and TA100. It is suggested that bamboo vinegars with a diluting percent content to the least extent of 20.00% or less have no cytotoxicity and mutagenicity, and their antimutagenic activity with NQNO and AFB_1 is effective. The bamboo vinegars in the extent of this report can be as the reference of functional additives, due to their prospective applications in human care, such as in the portion of drinks or/and food.

Author details

Han Chien Lin

Address all correspondence to: alexhlin@mail.ncyu.edu.tw

Department of Wood Based Materials and Design, College of Agriculture, National Chiayi University, Chiayi, Taiwan

References

[1] Austin R, Ueda K. Bamboo. New York: John Weatherhill Inc; p. 1970. 215 Record Number: 19780646836

[2] Forestry Bureau, Council of Agriculture, Executive Yuan. The Annual Statistics of the Agricultural Reports. Taiwan ROC: Forestry Bureau, Council of Agriculture, Executive Yuan Publication; 2015

[3] Wang D, Shen SJ. Bamboos of China. Portland, Oregon: Timer Press; 1987. p. 167

[4] Lin YZ, Hwang GS. Contribution of CO_2 reduction from wood based wastes charcoal materials. Forest Technology Report. 2006;**13**(1):23-27

[5] Hung CP. Application of wood/bamboo charcoal in life. Forest Technology Report. 2004;**11**(3):18-21

[6] Wang SY, Hung CP, Wang CK, Wan WT, Lai FT. Prosperity examination and activation preparation of bamboo. Industrial Materials. 2004;**205**:86-92

[7] Wang CK. Effects on the basic properties and ability of carbonization of Taiwan giant bamboo and bamboo carbonized materials [thesis]. Taiwan: National Taiwan University; 2004

[8] Lin YZ, Hwang GS, Wang IS. Introduction to production and utilization of bamboo charcoal. Forest Technology Report. 2003;**10**(3):31

[9] Lin HC, Li ST, Shiah TC, Duh MH, Fujimoto N. Charcoals evaluation regarding the effect of preservative-treated woods leached metal loss on water-based environment. Journal of the Faculty of Agriculture. Japan. 2016;**61**(2):351-359

[10] Lee BS. Bamboo charcoal for inside body environmental protection. All Health. 2007; **262**:69-70

[11] Yataga M, Nishimoto M, Hori K, Ohira T, Shibata A. Termiticidal activity of wood vinegar, its components and their homologues. Journal of Wood Science. 2002;**48**:338-342. DOI:10.1007/BF00831357

[12] Ikimoto T, Ikeshima O. Usages of Bamboo Charcoal and Bamboo Vinegar. Japan: Nobunkyou Publication; 2000. p. 320

[13] Nomura R. Healthy Effects of Bamboo Vinegar. Japan: Nobunkyou Publication; 2004. p. 245

[14] Uchimura T, Tanikai H, Hosoukawa K. Issues of Bamboo Charcoal and Bamboo Vinegar. Japan: Soumorisya Publication; 2000. p. 256

[15] Tsai CT. Evaluation of antioxidation activity and resist lipid oxidation effect with Moso bamboo vinegar collected from different temperatures [thesis]. Taiwan: National Chiayi University; 2009

[16] Lin HC, Ohuchi T, Murase Y, Shiah TC, Shieh SL, Chiu PC, Juan YM, Cheng SC. Application of bamboo vinegar with vacuum process to evaluate fungi resistance of bamboo materials. Journal of the Faculty of Agriculture. Japan. 2006;**51**(1):5-11

[17] Huang YP, Peng CW, Huang YH, Tsai TR, Kuo WR, Lin HC. Investigation of Mikania micrantha charcoal applied on cultural media. Materials Science Forum. 2011;**685**:216-229. DOI: 10.4028/www.scientific.net/MSF.685.216

[18] Lin HC, Huang YH, Tsai TR, Peng CW, Hwang GS, Kuo WR. Investigation of promoting growth of tomato plug seedlings from Moso bamboo charcoal cultural media with Bamboo vinegar. Journal of Agriculture and Forestry. (NCYU). 2011;**8**(2):58-72

[19] Ho HM, Kuo WR, Peng CW, Hwang GS, Lin HC. Investigation of yield and seedling growth of cabbage from cultural media with Moso bamboo charcoal. Journal of Agriculture and Forestry. (NCYU). 2013;**10**(1):43-60

[20] Hageta Y. Application of Wood/Bamboo Vinegar on the Cultivation of Grape. Japan: Nobunkyou Publication; 2004. p. 450

[21] Kobahasi Y. Reductive Effectives of Agricultural Chemical. Japan: Nobunkyou Publication; 2004. p. 580

[22] Yoshie L. Effects of Wood/Bamboo Vinegar on Cow Skin. Japan: Nobunkyou Publication; 2004. p. 520

[23] Kou CW. Effects of heating rates on fundamental properties of vinegar during bamboo carbonization and its application on inhibit plant pathogenic bacteria [thesis]. Taiwan: National Chung University; 2004

[24] Lin HC, Shiah TC. Evaluation of fungi resistance of Moso bamboo materials using bamboo vinegar with smoking process. Journal of Forestry Research. Taiwan. 2006;**28**(2):51-66

[25] Lu KT, Kuo CW, Liu CT. Inhibition efficiency of a mixed solution of bamboo vinegar and chitosan against ralstonia solanacearum. Taiwan Journal of Forest Science. 2007;**22**(3):329-338

[26] Lin HC, Murase Y, Shiah TC, Hwang GS, Chen PK, Wu WL. Application of Moso bamboo vinegar with different collection temperatures to evaluate fungi resistance of Moso bamboo materials. Journal of the Faculty of Agriculture, Kyushu University. Japan. 2008;**53**(1):107-113

[27] Chen PK, Wu SC, Hwang GS, Lin HC. Investigation on basic properties and antibacterial activity of bamboo/wood vinegars collected from different temperature. Journal of Agriculture and Forestry. (NCYU). 2010;**7**(2):95-117

[28] Namiki M. Antioxidants/antimutagens in foods. Critical Reviews in Food Science and Nutrition. 1990;**29**:273-300. DOI: 10.1080/10408399009527528

[29] Horne PA, Paul TW. Influence of temperature on the products from the flash pyrolysis of biomass. Fuel. 1996;**75**(9):1051-1059. DOI: 10.1016/0016-2361(96)00081-6

[30] Vasavada MN, Cornforth DP. Evaluation of antioxidant effects of raisin paste in cooked ground beef, pork, and chicken. Journal of Food Science. 2006;**71**(4):242-246. DOI: 10.1111/j.1750-3841.2006.00026.x

[31] Council of Agriculture Certified Agricultural Standards of Forest Products. Taiwan ROC: Council of Agriculture, Executive Yuan; 2004

[32] Ames BN, McCann J, Yamasaki E. Methods for detecting carcinogens and mutagens with the *Salmonella/Mammalian*-microsome mutagenicity test. Mutation Research. 1975; **31**:347-364. PMID: 768755

[33] McCann J, Choi E, Yamasaki E, Ames BN. Detection of carcinogens as mutagens in the *Salmonella*/microsome test: Assay of 300 chemicals. Proceedings of the National Academy of Sciences of the United States of America. 1975;**72**(12):5135-5139. PMCID: PMC388891

[34] Sugimura T, Sato S, Nagao M, Yahagi T, Matsushima T, Seino Y. Overlapping of carcinogens and mutagens. In: Magee PN, Takayama S, Sugimura T, Matsushima T, editors. Proceedings of the International Symposium of the Princess Takamatsu Cancer Research Fund, Tokyo, 1975. Fundamentals In: Cancer Prevention. vol. 6. Baltimore: University Par Press; 1976. pp. 191-215

[35] Ames BN, McCann J. Detection of carcinogens as mutagens in the *Salmonella*/microsome test: Assay of 300 chemicals. Proceedings of the National Academy of Sciences of the United States of America. 1976;**73**:950-954. PMID: PMC336038

[36] Waleh NS, Rapport SJ, Mortelmans KE. Development of a toxicity test to be coupled to the Ames *Salmonella* assay and the method of construction of the required strains. Mutation Research. 1982;**97**:247-256

[37] Mortelmans K, Zeiger E. The Ames *Salmonella*/microsome mutagenicity assay. Mutation Research. 2000;**455**:29-60. PMID: 11113466

[38] Chen PC. Evaluation of bamboo/wood vinegars collected from different temperatures on antibacterial activity and preliminary safety [thesis]. Taiwan: National Chiayi University; 2010

[39] Weng YC. Evaluation of adsorption properties and mutagenicity of activated carbons refined from charcoals [thesis]. Taiwan: National Chiayi University; 2010

[40] Maron DM, Ames BN. Revised methods for the *Salmonella* mutagenicity test. Mutation Research. 1983;**113**:173-215. PMID: 6341825

[41] Ministry of Health and Welfare. The Sanitation Standard for Edible Natural Colorants, Food Sanitation Standards. Taiwan ROC: Ministry of Health and Welfare, Executive Yuan; 1989

[42] Duh PD, Yen GC. Antioxidant activity of three herbal water extracts. Food Chemistry. 1997;**60**:639-645. DOI: 10.1016/S0308-8146(97)00049-6

[43] Yen GC, Duh PD, Hung YL. Contributions of major components to the antimutagenic effect of Hsian-Tsao (*Mesona procumbens* Hemsl.). Journal of Agricultural and Food Chemistry. 2001;**49**:5000-5004. DOI: 10.1021/jf0103929

[44] Huang YH. Antifungal activity of bamboo/wood vinegars collected from different temperatures against *Tinea pedis* fungi [thesis]. Taiwan: National Chiayi University; 2013

[45] Yam TS, Shah S, Hamilton-Miller JMT. Microbiology activity of whole and fractionated crude extracts of tea (*Camellia sinensis*), and of tea components. FEMS Microbiology Letters. 1997;**152**:169-174. DOI: 10.1111/j.1574-6968.1997.tb10424.x

Section 3

Characterization and Application

Chapter 5

Safe, Simple and Comfortable House with Bamboo Reinforced Concrete Structure

I Nyoman Sutarja

Additional information is available at the end of the chapter

http://dx.doi.org/10.5772/intechopen.68543

Abstract

Development of a livable housing in the form of a simple home is indispensable for poor families. Simple houses built with brick or concrete block walls and a frame of bamboo reinforced concrete structures are resistant to earthquake and wind as well as convenient. The construction method to consider the application of the six criteria for appropriate technology (technical, economical, ergonomic, social cultural, energy-saving, and environmentally friendly) and the systemic, holistic, interdisciplinary and participatory (SHIP) approach are applied starting from when a house is planned, constructed, and maintained. The conclusions of this research are as follows: (1) compressive strength of concrete, $f'c$ = 20.75 MPa and the tensile strength of bamboo fy = 129.17 Mpa; (2) the simple house with the bamboo reinforced concrete structure can fulfill the requirement of being safe prevailing in Indonesia; (3) it can fulfill the requirement of physical comfort, which includes the temperature, humidity, the wind speed, noise, and illumination; (4) based on the information obtained through the questionnaire, such a house can also make the dweller feel highly safe, comfortable, and satisfied.

Keywords: simple house, bamboo reinforced concrete, safe, comfortable

1. Introduction

Housing is one of the human basic needs apart from food and clothing. Everybody has the right to occupy a comfortable house. The reason is that house plays an essential role in human life. Without house humans cannot live comfortably as it importantly contributes to the daily human health, meaning that the indoor and outdoor human activities are also determined that their houses are comfortable [22].

A livable house should be safe, healthy, comfortable, and adequately equipped, as prescribed in the Act of Building Construction Number 28/2002 and the Regulation of the Ministry of Public Works Number 25/PRT/M/2007 dated on August 9, 2007 concerning "Pedoman Sertifikasi Laik Fungsi Bangunan Gedung" (Guideline for Functionally Comfortable Building Construction) [2].

It is highly possible for the natural disasters such as the earthquake and wind to take place in Bali as an island which is densely populated enough [4, 20]. It is recorded that the earthquake referred to as "Gejer Bali" which took place in 1815 cost 1500 lives; the earthquake referred to as "Gempa Buleleng" which took place in 1862; the Negara earthquake took place in 1890, the Gejer Bali earthquake occurred again in 1917 and cost 1500 lives, the Seririt earthquake occurred in 1976 and cost 560 lives, the Karangasem earthquake took place in 1979 and cost 24 lives. Several other earthquakes that were responsible for the material loss have also been recorded.

The bamboo has a high strain force and available almost everywhere in Indonesia, especially in Bali. Therefore, it is necessary to construct simple livable houses with bamboo reinforced concrete structure, which are resistant to the earthquake and wind. The reason is that they can be constructed using the cheap local building materials [5]. In addition, such houses are functionally feasible. However, it was necessary to test them to identify how reliable and comfortable they were and to what extent they could satisfy the dwellers. The test was done by building two types of simple houses; one is the simple traditional Bali age house and the simple modern Balinese house.

The six criteria of the appropriate technology (technical, economic, ergonomic, socio cultural, energy saving, and environmentally friendly) with the systemic, holistic, interdisciplinary and participatory (SHIP) approach are relevantly applied starting from when a house is planned, constructed, and maintained. The design of a house should develop and grow based on what is needed by people; it should not be designed by other parties with particular objectives [1]. The designer should pay attention to the human dimensions (the human ability and limitedness) [25]. A house should be designed based on the user; it should be constructed in such a way that it can reduce the possible negative impact and improve the user's comfort. If it is designed based on what is needed and expected by the user, it can improve the quality of his/her life [6].

As times go by, an idea has appeared recently that the local wisdom should be applied to overcoming the human problem and environment. The reason is that it is easily accepted by the local people. Exemplifies [18] the concept *Tri Hita Karana* especially the relationship between the user of a house with his/her ancestors and the relationship between him/her with his/her environment, the concept *Tri Mandala* which suggests that a compound should be divided into three zones: they are the highest "hulu" zone, the middle "tengah" zone, and the lowest "teben" zone; the concept *Tri Angga*; the house should be physically divided into three parts: they are the roof, the body (wall), and the foot (foundation).

2. Material and method

2.1. Material and research location

The test was done by building two types of simple houses: they are the simple traditional Bali age house and the simple modern one. The study was conducted in Bali Province, Indonesia, where different types of bamboo can grow which can be used as concrete reinforcement. This present study gives a new opportunity that bamboo can be used as the reinforced concrete structure of the simple house resistant to the earthquake and wind.

2.1.1. *The brick masonry*

The brick masonry tested was made at Keramas village, Gianyar regency (one of the villages close to Pengotan village). The mortar space was made of Portland cement and sand with the proportion one to four. The testing process is described in **Figure 1** [23].

2.1.2. *Result of the bamboo tensile strength test*

The bamboo tensile strength was obtained using the procedure regulated in the prevailing material tensile strength test in the laboratory, as shown in **Figure 2** [22]

2.1.3. *The nominal moment and ultimate moment*

Based on the concrete compressive strength and bamboo tensile strength, the nominal moment of the beam could be analyzed. Beam dimensions were 110 mm width and 250 mm height. The bamboo reinforcement area of 400 mm^2 was used for beam 1a, 1b, 1c, and 1d while for beam 2a, 2b, 2c, and 2d was used bamboo reinforcement area of 1200 mm^2. **Figure 3** [22] shows the process of how the bamboo reinforcement was assembled. **Figure 4** [22] shows the loading test of the bamboo reinforced concrete beam.

2.2. Research design

This present study is an experimental one with the treatment by subject design. This method was chosen in order to be able to analyze the four factors of life quality, such as safety, thermal comfort, subjective comfort, and the life satisfaction of the user, after the simple houses with the bamboo reinforced concrete structure were constructed.

Figure 1. The brick masonry compressive strength test.

Figure 2. Standard machine used to test the bamboo tensile strength [22].

Figure 3. The bamboo reinforced concrete beam assembled was 120 mm × 250 mm. The reinforcement area were 400 and 1200 mm² [22].

Figure 4. The loading test of concrete beam which 120 mm × 250 mm with the reinforcement area were 400 and 1200 mm² [22].

2.3. The house construction

The diagram of the house construction to which the six criteria of the appropriate technology with the systemic, holistic, interdisciplinary and participatory (SHIP) approach is applied can be described as follows (**Figure 5**) [23]. Such a house construction gives emphasis on the local wisdom-based people's participation.

The simple house construction using the people's participatory approach can definitely and gradually improve the people's capacity due to the following reasons:

1. Promoting job opportunities: opportunities are made available for people to participate in developing their economy justly and equally, meaning that poverty can be reduced.
2. Community empowerment: the quality of human resources becomes improved, meaning that people can have access to and participate in the decision making.
3. Capacity building: the poor people can increase their income by improving their health, education, skill, technology, and information using the on the job training (OJT) method.
4. Social protection: the poor people become protected and feel safe from the natural disaster, social conflict, and so forth.

The simple house construction using the participatory approach takes the following principles into consideration:

1. The top-down way of thinking changes into the bottom-up way of thinking
2. Priority is given to the simple technology
3. The user is made to have the sense of belonging and maintaining
4. Technologically, people are made to improve their technical and administrative expertise.
5. No environmental degradation takes place.

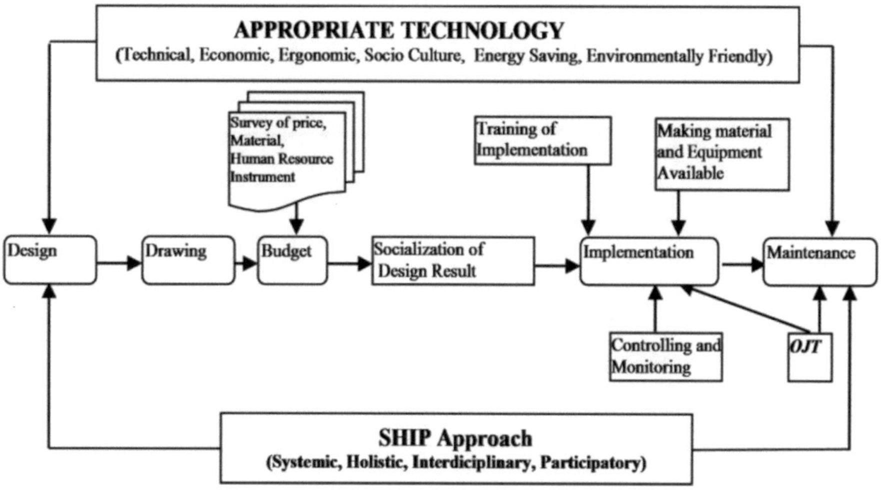

Figure 5. Diagram of the house construction using the participatory approach [23]. Note: OJT = on the job training.

2.4. Data analysis

The data analyzed to determine whether the simple houses were livable or not are as follows:

1. The building physical data: the dimension and the material strength were analyzed to know whether the houses were resistant to the earthquake and wind or not. The finite element method assisted with the SAP-2000 program.
2. The level of thermal and physical feasibility: it was measured using the environment meter (reliable testing & measuring equipment—KRISBOW-kw06-291).
3. The extent to which the dweller feels satisfied and subjective comfortable was measured through questionnaire.

3. Result and discussion

3.1. Characteristic of the construction material

For, the construction of simple house with either brick or concrete block and the bamboo reinforced concrete structure characteristics of the materials are need be determined. To this end, the characteristics of the materials used were tested in the laboratory. The material tests are as follows: compressive strength of brick masonry, compressive strength of concrete block wall, compressive strength of concrete, tensile strength of bamboo, and flexural test on concrete beam with bamboo reinforcement.

3.1.1. The brick masonry compressive strength

Figure 1 shows the four samples of brick masonry used in the test: the length averaged 470 mm, width averaged 112 mm, and the height averaged 377 mm. The average compressive strength obtained was 2.04 MPa.

3.1.2. The concrete block wall compressive strength

The mortar space of concrete block wall was made of Portland cement and sand with the ratio 1 to 4. The compressive strength averaged 4.96 Mpa, which, according to SNI 03-0348-1989, can be classified as having the quality of B40. **Figure 6** [3] shows a failure/collapse model of

Figure 6. Collapse of concrete block wall sample after being tested [3].

concrete block for wall system in Indonesia under compressive test. The concrete block wall used to apply as a typical masonry wall in semimodern residential houses.

3.1.3. Result of the bamboo tensile strength test

The bamboo tensile strength was obtained at the laboratory of the Mechanical Engineering Department of Udayana University, as shown in **Figure 7** [22] and **Table 1** [22].

It can be seen from **Table 1** that the average yield strength was 129.17 MPa and the average elongation was 8.99%. The results were used to analyze the nominal flexural strength or the nominal moment of the bamboo reinforced concrete. Viewed from the tensile strength, it can be stated that bamboo can be used as an alternative concrete reinforcement, especially the light and moderate structure.

3.1.4. The result of compressive strength of concrete cylinder

Three concrete cylinders were tested in order to identity the concrete strength. The diameter of each cylinder was 150 mm and the height was 300 mm. The results were that the compressive

Figure 7. The bamboo was tested until it was Fracture [21].

0.2 Y.S.	Yield strength	Rupture strength	Elongation
(MPa)	(MPa)	(MPa)	(%)
130.67	130.67	180.00	8.865
127.67	127.67	201.00	9.128

Table 1. The result of the bamboo tensile strength test.

strength of cylinder 1 was 21.07 MPa, cylinder 2 was 20.35 MPa, and cylinder 3 was 21.94 MPa, meaning that the average compressive strength was 21.12 MPa.

3.1.5. The nominal moment and ultimate moment

Figure 8 shows the crack pattern of the concrete beam after being tested [21].

Table 2 [22] presents the result of the nominal moment analysis of the beam and ultimate moment analysis. All the blocks were collapsed beyond to the nominal strength; the average ultimate strength was 22.46% greater than the nominal strength.

3.1.6. The structural system of bamboo reinforced concrete

The Bali age traditional house and the modern Balinese house constructed for this study used the structural system of bamboo reinforced concrete. The bamboo was splitted in dimension 10 mm × 30 mm. The bamboo was tied to steel stirrup with an 8-mm diameter, as shown in **Figure 9** [24]. The walls of the traditional Bali age house were made of brick masonry which spacing made of mortar with proportion of 1 Portland cement to 4 sand, whereas the walls of the modern Balinese house were made of concrete block.

3.2. The traditional Bali age house

3.2.1. The characteristics of the house used as the sample

The characteristics of the Bali age traditional house used as the sample in this present study included the total area of the building, the area of the room, the area of the terrace, the height

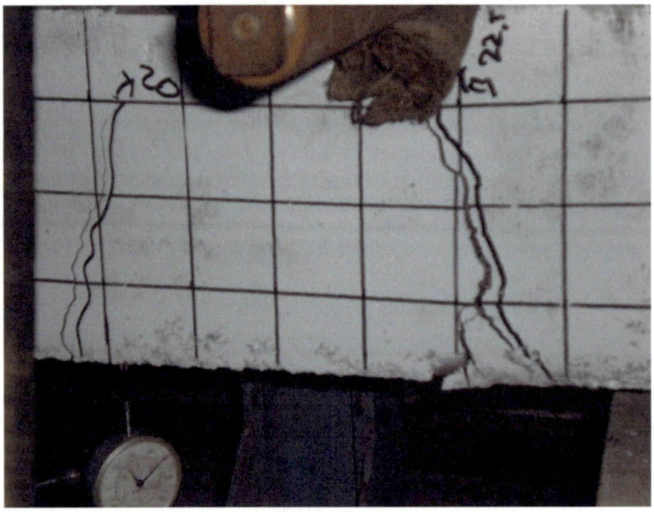

Figure 8. The crack pattern of the bamboo reinforced concrete beam after being tested [22].

No.	Number of tested items	Nominal moment (kNm)	Ultimate moment (kNm)
1	1a	10.47	13.2
2	1b	10.47	13.0
3	1c	10.47	10.8
4	1d	10.47	19.2
5	2a	27.42	29.2
6	2b	27.42	35.2
7	2c	27.42	30.4
8	2d	27.42	25.2

Note: The nominal moment was analyzed based on the formulation proposed by Ghavani [5].

Table 2. Presents the nominal moment and the ultimate moment.

Figure 9. The bamboo reinforcement was assembled before concrete was casted [24].

of the walls, the number, height, and width of the doors. Four units of the traditional house were used as the sample in the present study with the characteristics presented in **Table 3**. The componential dimension of one unit was different from that of another, as it should be adjusted to the dweller's anthropometry. **Figure 10** [23] shows a residential house-1 built by using brick wall and bamboo reinforced concrete structure system.

In order to find out the displacement of the top of the building under seismic load, traditional residential houses were analyzed by using three-dimensional configuration, as shown in **Figure 11** [23]. From the result of the three-dimensional analysis of the four traditional houses, the average maximum horizontal displacement was 0.193 cm with the building height was 3750 cm, thus displacement ratio was 0.0005 or 0.05%.

Characteristics	House-1	House-2	House-3	House-4
The building area (m^2)	36.00	40.95	29.00	22.50
The room area (m^2)	25.80	30.87	20.30	15.00
The terrace area (m^2)	10.20	9.10	8.70	7.50
The door height (m')	1.95	2.20	2.10	1.95
The number of doors	1	1	1	1
The door size (W × H)	1.80 × 0.70	1.85 × 0.70	1.75 × 0.70	1.70 × 0.70
The number of widows	2	2	2	2
The door size (W × H)	1	1	1	1
The roof material	Tile	Tile	Tile	Tile
The walls material	Brick	Brick	Brick	Brick
The floors material	Cement	Cement	Cement	Cement
The terrace floors material	Ceramic	Ceramic	Ceramic	Ceramic

Table 3. The characteristics of the traditional Bali age house.

Figure 10. The simple traditional Bali age house with the bamboo reinforced concrete structure and brick walls [23].

3.2.2. *Thermal comfort*

In this present study, the environmental condition included the dry temperature, the relative humidity (RH), the wind speed, illumination, and noise, which contributes to the physical comfort. The environmental condition, after and before the internal and external parts of the house were redesigned, is presented in **Table 4**.

http://dx.doi.org/10.5772/intechopen.68543

Figure 11. Three-dimensional model of structural system (example of model house 1 [23].

Nbr.	Description	Time when it was measured			
		06.00	**12.00**	**18.00**	**24.00**
1	The temperature in the room (°C)	21	27	24	22
2	Humidity (%)	78	72	79	82
3	Noise (dBA)	39	42	40	35
4	Natural illumination (Lux)	135	245	215	–
5	The wind speed (m/second)	0.1	0.3	0.2	–

Table 4. The physical comfort of the traditional Bali age house.

The temperature within the room ranged from 21 to 27°C or it averaged 23.50 ± 2.65°C. The relative humidity (RH) within the room averaged 77.75 ± 4.19%. The maximum noise during daytime was 42 dBA and at night it was 35 dBA, and averaged 39.00 dBA. The average natural illumination within the room during daytime ranged from 135 to 245 Lux, higher than 115 Lux, the minimum natural illumination required for doing activities within the bed room [7]. The wind speed within the room during daytime ranged from 0.1 to 0.3 m/second. The natural illumination and wind speed within the room were not measured as the window and ventilation were closed.

3.3. The modern Balinese house

3.3.1. The characteristics of the house used as the sample

The characteristics of the modern house used as the sample in this present study included the whole area of the house, the area of the room, the area of the terrace, the height of the wall, the number, height and width of the door. All the componential dimensions were adjusted to the dweller's anthropometry as presented in **Tables 5** and **6** [24]. **Figure 12** [24] shows a modern Balinese house which built of concrete block wall with a bamboo-reinforced concrete system.

3.3.2. Structural performance of the house

The structural performance is defined as a ratio of the most maximum horizontal displacement to the height of the building. The building was analyzed in three-dimensional model as

Nbr.	The house characteristics	Volume	Unit
1	Area of the building	35	m^2
2	Area of the room	2 × 12.5	m^2
3	Area of the terrace	1 × 10.5	m^2
4	Height of the wall	3.1	m'
5	Number of the doors	2	pieces
6	Number of the windows	2	pieces
7	Foundation material	Stone	–
8	Wall material	Concrete block	–
9	The hood	Hood	–
10	Roof material	Pressed tile	–
11	Floor material	Ceramic	–

Table 5. The house characteristics.

Nbr.	Description	Width (cm)	Height (cm)	Reinforcement area (cm^2)
1	Sloof (concrete)	20	25	8.40
2	Column (concrete)	20	20	11.20
3	Beam (concrete)	18	30	7.00 + 2.80
4	Gording (wood)	6	12	–
5	Rafter	5	6	–
6	Lisplank (wood)	2	20	–

Table 6. The characteristics of the house structural component.

the consequence of the load assigned. After that the maximum horizontal displacement could be determined.

Figure 13 [24] shows a typically modern three-dimensional model of simple Balinese houses which has been analyzed to define deformation at the top of the structures under earthquake load. From the result of the three-dimensional analysis of the house, the maximum horizontal displacement was 1.71 cm with the building height was 5500 cm, thus the displacement ratio was 0.031%.

3.3.3. The building physical comfort

The measured building physical comfort included the temperature, humidity, noise, and the natural illumination. It was measured at 6 a.m., 12 a.m., 6 p.m., and 12 p.m. The results of the measurement are presented in **Table 7** [24].

Figure 12. The simple modern Balinese house with the bamboo reinforced concrete structure and concrete block wall [24].

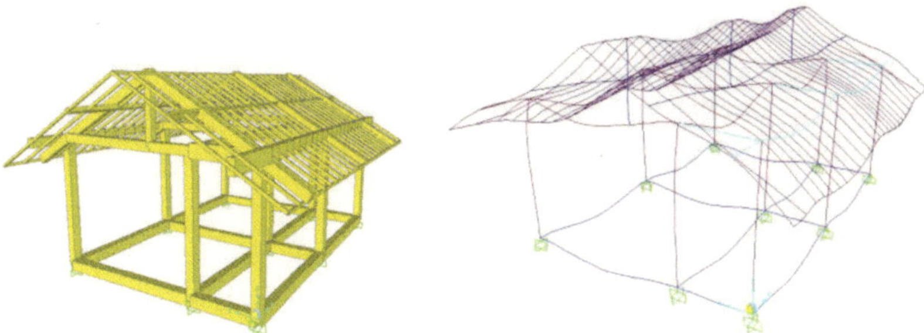

Figure 13. Model and deformation of house structure [24].

Nbr.	Description	Time of measurement			
		06 a.m.	12.00 a.m.	06 p.m.	12.p.m.
1	The temperature within the room (°C)	26	32	31	28
2	Humidity (%)	79	67	77	80
3	Noise (dBA)	41	41.8	44.74	41.47
4	Natural illumination (Lux)	145	255	235	–
5	Speed of wind (m/second)	0.1	0.4	0.25	–

Table 7. The physical comfort of the modern Balinese house.

3.4. Discussion

3.4.1. Construction material

The construction forming materials tested in the laboratory included the compressive strength of the concrete block, the compressive strength of the brick, the compressive strength of the concrete, and the tensile strength of the bamboo. The materials were tested as the data used to analyze the performance of the house structure. The results showed that the compressive strength of the concrete block averaged 4.96 MPa and the compressive strength of the brick averaged 2.04 MPa. The compressive strength of the concrete cylinder with a diameter of 150 mm and a height of 300 mm averaged 21.12 MPa or equal to K-260 kg/cm^2, meaning that the standard SNI – 213 for concrete was already fulfilled which requires that the minimum strength of the concrete structure in an area which is sensitive to the earthquake such as Bali should be 20.75 MPa and equal to K-250 kg/cm^2.

Bamboo test results show in **Table 1**, in which the tensile strength of the bamboo averaged 129.17 MPa and the rupture strength averaged 191 MPa, so the over strength was 47.86%. The melting and tensile strength fy = 129 MPa was used to design and analyze the diameter of the item tested. Bamboo can be used as the alternative reinforced concrete structure in the construction of a simple house. In addition, it is cheap and can be found anywhere in Bali. This is supported by the result of the study conducted in Ref. [5], stating that bamboo can be used as an alternative material of construction because it is cheap and saves energy. According to Ref. [9], the tensile strength of the bamboo ranges from 200 to 300 MPa, its flexibility averages 84 MPa and its elastic modulus is 200.000 MPa. Viewed from its tensile strength, it is feasible enough to use as the concrete reinforcement, at least for the light and medium structure. Furthermore, according to Ref. [14], bamboo can be used as the reinforcement of the concrete structure as its rupture strength is high enough [16].

3.4.2. Safety and performance of construction

Based on the compressive strength of the concrete and the tensile strength of the bamboo, it can be stated that the bamboo-reinforced concrete structure with brick walls used to construct the traditional Bali age house is resistant enough to load assigned (dead load, live load, earthquake load, and wind load), and bamboo reinforced concrete structure could be used to construct the simple modern Balinese house. It can be stated that the two types of the houses are safe enough for the dwellers. The bamboo reinforced concrete structure was already applied to constructing two storied houses. It was stated in the study conducted by Virgyan [26] that the multiple bamboo reinforced portal concrete structure that was bridled at the construction location with plastic joints can be used in the construction of a simple house, which is resistant to the earthquake. Based on the result of the three-dimensional analysis, it can be stated that the performance of the structural system is highly good, making the dwellers feel so safe. The performance of the structural system when planning to construct a building, which is resistant to the earthquake in Indonesia, is highly important as it is highly possible for the earthquake to take place in almost every part of Indonesia [28].

3.4.3. *Thermal comfort*

The thermal comfort or the house physical environment is affected by the temperature, relative humidity, speed of the wind, illumination, and noise. The temperature within the room in the traditional Bali age house ranges from 21 to 27°C, and the temperature within the room in the modern Balinese house ranges from 26 to 32°C. At certain hours, it is higher than what is required. In the equatorial region, the comfortable temperature ranges from 22.5 to 29.5°C [12], from 21.37 to 28.37°C (ASHRAE), and from 22.8 to 30.2°C [21].

The relative humidity both within and outside the room is almost the same, namely 60% during daytime and 80% at night. It will be better if the relative humidity is higher than 20% all the year round, and lower than 60% during the summer and lower than 80% during the winter [11]. If the relative humidity is higher than 80%, there will be water vapor on the human skin, making the body uncomfortable [17, 19]. In addition, the dweller will not be in good health and, for example, there will be fungus on the skin.

The speed of the wind within the room ranges from 0.1 to 0.4 m/second during daytime. At night, it is 0 m/second as the window and ventilation are closed. The speed of the wind can contribute to the speed of the missing heat due to convection and evaporation. Therefore, the speed of the wind ranging from 0.1 to 0.3 m/second can fulfill what is required for being comfortable [11, 13]. It should not be faster than 0.2 m/second [10]. As the speed of the wind is in accordance with what is required by the experts mentioned above, the weather within the room circulates well, causing the dweller's health to improve and the eyes to be less irritated. The front window and the ventilation at the rear wall affect the speed of the wind and cause the cross-weather circulation to take place. This is supported by the result of the study conducted in Ref. [15], in which it is stated that the cross-weather circulation can improve the comfort of those who stay at the simple houses located in Cemara Giri area, Dalung, Bali.

The natural illumination within the room ranges from 145 to 255 Lux during daytime, higher than what is required, namely 115 Lux [27]. The intensity of the natural illumination within the room is affected by how wide the window is open and how wide the ventilation is. The window or ventilation is installed at the front and rear walls, causing the fresh weather to circulate from the front window and the dirty weather to get out from the rear ventilation. The dweller becomes comfortable and the electric energy can be saved. The maximum use of the natural illumination does not only positively contribute to the dweller's health but also saves the electric energy. Before the house was redesigned, the electric lamp was always turned on when there was an activity within the room. The good natural illumination can make the room brighter and healthier. Indra [8] stated that the maximum penetration of the sunlight into the house reduce the lamp use, causing the costs needed to be reduced. It is proved that the sunlight can also kill the bacteria which can grow well in the humid environment. Vitamin D which is obtained from the sunlight is highly useful to the bones and skin [27].

The maximum noise within the room is 44.74 during daytime and 41.47 dBA at night. It is still under the limit determined by the government of Bali province, namely 50 dBA during daytime and 45 dBA at night. The noise taking place in the residential environment is more frequently caused by the agricultural equipment such as tractor. The acoustic safety such as noise cannot be separated from the dweller's health; however, as its effect cannot be identified at once, the impact cannot be felt at once either. Therefore, the acoustic safety is still neglected. The dweller does not only feel uncomfortable but his/her health will also be indirectly getting worse. As an illustration, he/she cannot take a rest well due to noise. As a consequence, he/she will get tired, angry, cannot concentrate well, and so forth. If this takes place continuously, it is highly possible that his/her health will be getting worse.

3.4.4. The dweller's safety, comfort, and satisfaction

The houses that had been constructed were directly occupied by the dwellers. Their subjective response to safety, comfort, and satisfaction was measured using questionnaire. It was measured after the houses were occupied for 30 days. On the average, they felt highly comfortable, safe, and satisfied. The participatory approach to the construction caused the dwellers to feel highly satisfied. They were directly involved from the planning to the construction.

3.4.5. The dweller's life quality

The people's dynamic human life always interacts with the environment where they live, the facilities and infrastructure they use, the organization they belong to, the activities they do and the houses where they can take a rest. The house that is constructed using the concrete block as the walls with the bamboo reinforced concrete structure can make the environmental condition safe and comfortable for its dwellers. This can improve their health quality as they can do their activities in the healthy, comfortable, and safe environment [1].

The anthropometry that is adjusted to the facilities and infrastructure available leads to an easier access [25]. Similarly, the maximum natural illumination contributes to the dweller's safety. It saves the electric energy that then can reduce the household costs. In addition, it can also hamper the global warming and delay the natural sources such as the coal from being used up [8].

The SHIP approach, which gives emphasis on the dweller's participation from the planning, during the construction and in the maintenance of the house can make the dweller more satisfied, as what he/she suggests can be accommodated [1]. In short, the simple house with the bamboo reinforced concrete structure fulfills the requirements of being healthy and safe and this positively contributes to the dweller's life quality. If a house is constructed based on what is needed and expected by the dweller, it can function as a means of improving his/her life quality [6].

4. Conclusion and suggestion

Based on the results of the study, analysis and discussion, several conclusions can be drawn as follows:

1. The simple house with the bamboo reinforced concrete structure can fulfill the requirement of being safe prevailing in Indonesia.
2. It can fulfill the requirement of physical comfort, which includes the temperature, humidity, the wind speed, noise, and illumination.
3. Based on the information obtained through the questionnaire, such a house can also make the dweller feel highly safe, comfortable, and satisfied.

This is suggested that the bamboo reinforced concrete structure should be taken into consideration in the simple house construction.

Author details

I Nyoman Sutarja

Address all correspondence to: nsutarja_10@yahoo.com

Civil Engineering Department, Faculty of Engineering, Udayana University, Bali, Indonesia

References

[1] Adiputra N, dan Manuaba A. Adaptasi Rumah Adat Di Desa Penglipuran. Dinamika Kebudayaan II (3), Pusat Penelitian Kebudayaan dan Kepariwisataan Universitas Udayana. Denpasar: Lembaga Penelitian Universitas Udayana. 2000

[2] Anonim. Peraturan Menteri Pekerjaan Umum, Nomor 25/PRT/M/2007, 9 Agustus. Jakarta. 2007

[3] Anggreni M, Sudarsana IK, Sukrawa. Perilaku Tekan dan Lentur Dinding Pasangan, Batako Tanpa Plesteran, Dengan Plesteran dan Dengan Perkuatan Wiremess. Jurnal Spektran. Denpasar. 2015;**3**(2), (hal 10-19)

[4] Endro T. Bali Rawan Gempa Bumi, Denpasar: Tokoh, 21-27 November (14), 2010

[5] Ghavani K. Bamboo: Low Cost and Energy Saving Construction Materials, Modern Bamboo Structures. London: Taylor & Francis Group; 2008

[6] Hartatik PS, dan Sri Nastiti NN. Peningkatan Kualitas Hidup Penghuni di Rusunawa Urip Sumoharjo Pasca Redevelopment. Surabaya: disampaikan dalam Seminar Nasional Perumahan Permukiman Dalam Pembangunan Kota; 2010

[7] Hindarto P. Memaksimalkan Pencahayaan Alami Dalam Rumah Kita [Internet]. 2011 Available from: http://astudioarchitect.com/2011/04/memaksimalkan-pencahayaan-alami-dalam.html. [Accessed: 01-01-2012]

[8] Indra ZP. Pencahayaan Alami Solusi Utama Mendapatkan Kenyamanan dan Penghematan Listrik [Internet]. 2011 Available from: http://www.ideaonline.co.id/iDEA/Tips/Dekorasi-ruang/Ruang-transisi-dan-servis/Area-Servis-Unik-dengan-Pencahayaan-Alami [Accessed: 02-01-2012]

[9] Janseen JJ. The Mechanical Properties of Bamboo used in Construction. Canada: IDRC; 1987

[10] Kroemer KHE, Grandjean E. Fitting The Task To The Man, A Textbook of Occupational Ergonomics. 8th ed. London: Taylor & Francis; 2000

[11] Lechner N. Heating, Cooling, Lighting, Design Methods for Architects. John Wiley & Sons Ins; 2007. ISBN 978-979-769-127-1

[12] Lippsmeier G. Bangunan Tropis. Jakarta: Erlangga. 1994

[13] Mangunwijaya YB. Pasal-pasal Penghantar Fisika Bangunan. Jakarta: Penerbit PT. Gramedia; 1981

[14] Morisco. Bambu Sebagai Bahan Rekayasa. Yogyakarta: Fakultas Teknik UGM; 1996

[15] Nitiyasa NPN. Aspek Kesehatan dan Kenyamanan Perumahan Sederhana di Kawasan Cemara Giri Dalung Bali. Jurnal Teknik Gradien. 2009;**1**(1)

[16] Pheasant S. Ergonomics, Work and Health. Hong Kong: Macmillan Press Scientific & Medical

[17] Rahman, A. Identifikasi Kenyamanan Thermal Webb pada Rumah Tinggal di Tanah Lahan Basah, Banjarmasin: Prosiding PPI Standarisasi; 2010, 4 Agustus

[18] Rumawan PS. Peran Kearifan Lokal dalam Penataan Ruang di Bali, Kearifan Lokal dalam Pengelolaan Lingkungan Hidup, Denpasar: UPT. Penerbit Universitas Udayana; 2007

[19] Satwiko. Fisika Bangunan, Edisi Pertama, Yogyakarta: Andi; 2009. ISBN: 978-979-29-0734-6

[20] Subandono. Mitigasi Bencana Gempa Bumi dan Tsunami, Direktorat Pesisir dan Lautan, Ditjen Kelautan, Pesisir dan Pulau-Pulau Kecil, Departemen Kelautan dan Perikanan; 2008

[21] Sujatmiko W. 2011, Kenyamanan Termal Adaptif Pada Bangunan Perkantoran Yang Memiliki Tingkat Infiltrasi Udara Luar Cukup Tinggi Sebagai Basis Data Menuju Standar Kenyamanan Termal Indonesia, Prosiding PPI Standardisasi 2011 – Yogyakarta, 14 Juli 2011 (149)

[22] Sutarja IN. Kuat Lentur Balok Beton Bertulang Bambu, Denpasar: Universitas Udayana; 2012

[23] Sutarja IN. The application of brick wall and bamboo reinforced frame for redevelopment of traditional houses in Pengotan Village. Key Engineering Materials. 2014;**594-595**:477-481. Trans Tech Publication, Switzerland. DOI: 10-4028/www.scientific.net/KEM.594-595.477

[24] Sutarja IN. Kelayakan Rumah Sederhana Dengan Sistem Struktur Beton Bertulang Bambu, Penelitian Hibah Unggulan Universitas Udayana. 2016

[25] Sutjana IDP. Desain Produk dan Resikonya [Internet]. 2010. Available from: http://ejurnal.unud.ac.id, [Accessed: 09-10-2010]

[26] Virgyan AP. Kapasitas Gaya Geser Maksimum Model Struktur Portal Bertulang Bambu System Ganda dikekang dilokasi Sendi Plastis [Internet]. 2009. Available from: http://library.um.ac.id/free-contents/printjournal.php/42445.html [Accessed: January 1, 2012]

[27] Wibisono M. Pencahayaan Dalam Desain Rumah Tinggal, Tren Arsitektur 2010 [Internet]. 2010 Available from: http://mtmdesign.wordpress.com/2010/08/10/, [Accessed: January 1, 2011]

[28] Yosafat. AP. Evaluasi Kenerja Gedung Beton bertulang Tahan Gempa dengan Pushover Analysis, sesuai ATC-40, Fema 356 dan Fema 440, Bandung. Jurnal Teknik Sipil. 2006;**3**(1). Januari, Jurusan Teknik Sipil, Universitas Kristen Maranatha

Chapter 6

Bamboo Wear and Its Application in Friction Material

Yunhai Ma, Yucheng Liu and Jin Tong

Additional information is available at the end of the chapter

http://dx.doi.org/10.5772/intechopen.69893

Abstract

Sliding wear behaviour of bamboo (*Phyllostachys pubescens*) was investigated in the cases of dry friction. The wear volume of bamboo was a function of the sliding velocity, the normal load and the relative orientation of bamboo fibres with respect to the friction surface. And tribological properties of the Bamboo Fiber Reinforced Friction Materials (BFRFMs) were tested on a constant speed friction tester. The results showed that the wear volume increased with the increase of sliding velocity and normal load. The normal-oriented specimens (N-type) showed sound wear resistance in comparison to the parallel-oriented ones (P_S- and P_I-type), and the outside surface layer (P_S-type) showed sound resistance in comparison to the inner later (P_I-typ). The friction coefficient of BFRFMs (reinforced with 3 wt.%, 6 wt.% and 9 wt.% bamboo fibers) were higher than those of the non-bamboo fiber reinforced friction material with identical ingredients mixed with and process conditions during the temperature-increasing procedure. The friction coefficients of the specimens containing 3 wt.% bamboo fibers were higher than that of other specimens. The wear rate of BFRFMs increased with the increasing of test temperature, and the wear rates of specimens containing 3 wt.% bamboo fibers were lower than that of others specimens.

Keywords: bamboo, sliding wear, bamboo fiber, friction material, friction, wear

1. Introduction

Bamboo is a natural biomaterial and consists of vascular bundles (cellulose fibers). The vascular bundles are composed of many right-handed spiral phloem fibers at a certain spiral angle. Lakkad and Patel [1] investigated the mechanical properties of bamboo specimen, such as, Young's modulus, tensile strength, compressive strength, and interlaminar shear along fibers and the tensile strength across fibers. Bamboo has higher specific tensile strength than glass-reinforced plastic (GRP) and mild steel with chopped strand mat and woven roving, and comparable specific modulus with mild steel and GRP. Godbole and Lakkad [2] studied the influences of water absorption on mechanical performances of bamboo. The tensile strength, compressive strength, tensile modulus, and interlaminar shear of bamboo specimen reduced after soaking or boiling in distilled water. Li et al. [3] designed a double-fold spiral bionic composite model imitating the characteristic structure of bamboo. The tensile strength of carbon fiber reinforced tin composites processed by the bionic model was higher by 40% than that of unidirectional carbon fiber reinforced ones.

Yakou and Sakamoto [4] investigated abrasive performances of bamboo specimen with carborundum paper as the counterface. They indicated that the abrasive wear rate of the inner layer was higher than that of the outside surface layer for bamboo specimens of normal- and parallel-oriented cellulose fibers relative to the friction surface. Tong et al. [5] evaluated the abrasive wear properties of bamboo specimens by using quartz sand particles as abrasive material. The results showed that the abrasion resistance of bamboo specimen was decided by the relative orientation of the cellulose fibers with respect to the friction surface and by the size of abrasive particles. The abrasive wear rate increased with the increase of size of abrasive particles. Specimens with the normal orientation of cellulose fibers to the friction surface presented better abrasion resistance than those with the parallel orientation; the inner layer had lower abrasion resistance than the outside layer; and the cellulose fibers had better resistance than the matrix tissue. The dry sliding wear behavior of bamboo was studied in order to obtain some useful information for designs of friction materials.

On the other hand, friction materials are the key parts of automobiles brake systems, and many studies have been investigated to improve brake properties in order to adapt the people's requirement for security and rapid development of automobile [6]. To acquire comfortable and dependable brake properties of the automobiles, braking friction materials usually contain more than 10 different components. The components are normally classified as reinforced fibers, binders, property modifiers, and fillers. Each component plays an important role for brake performance under different braking conditions. Many studies investigated the effect of different components on brake performance [7, 8]. A related review on frontiers of fundamental tribological research emphasized the concern over the environmental protection, for instance, biodegradability in the development of tribo-materials [9].

Asbestos fibers, which have been widely used in braking friction materials, are harmful to human health and environment; they have been forbidden to be used for manufacturing friction materials. Therefore, substitutes of the asbestos fibers, such as steel fibers, Al_2O_3 fibers, carbon fibers, glass fibers, aramid fibers, copper fibers, and their hybrid fibers [10–12] have been studied and selected. Moreover, many study results showed that these fibers have excellent properties for friction materials; however, there are many shortcomings (such as weak combination strength, high cost, and high noise) to be resolved when these fibers are applied in friction materials.

Many researches have focused on the utilization of biological fibers with the function of protecting environment, such as betel nut fibers [13], cotton fibers [14, 15], jute fibers [16–18], kenaf and ramie fibers [19], sisal and flax fibers [20–22], and sugarcane fibers [23]. More and more biological fiber reinforced composite materials have been used as tribological ones. Dimensions and orientation of fibers in friction materials are important factors affecting its tribological properties. Such natural biomaterials with composite structures as bamboo provide clues and ideas for designs of friction materials, for example, antifriction materials and wear-resistant materials. The excellent structure of bamboo fiber was investigated by Paramesaran and Liese [24]. Bamboo fibers were used as one of the components in friction materials because of their low density and excellent mechanical properties in comparison to that of glass fibers [25–27]. Meanwhile, the bamboo fiber is a kind of plant fiber with cellulose structure and vascular bundles consisting of the fiber bundle and sclerenchyma sheaths.

In the present study, sliding wear behavior of bamboo (*Phyllostachys pubescens*) against a gray iron (HT200) was investigated in the cases of dry friction. The wear volume of bamboo was a function of the sliding velocity, the normal load, and the relative orientation of bamboo fibers with respect to the friction surface. Moreover, the mechanical and physical properties of the bamboo fibers were studied. Comparative studies were performed to investigate the effects of bamboo fiber content on the friction performances of friction materials. The tribological property of the friction materials was evaluated and discussed at the test temperature of 100–350°C. The wear surface morphologies and the wear mechanism of bamboo fiber reinforced friction materials (BFRFMs) were analyzed using scanning electron microscopy (SEM).

2. Experimental materials and methods

2.1. Bamboo wear experiments

Bamboo specimens of dimensions 14 × 10 × 8 mm were cut from the air-dry bamboo (*P. pubescens*). Three types of specimens were prepared. The friction surface was normal to the cellulose fiber orientation for the N-type, and parallel to both the cellulose fiber orientation and the outside surface of bamboo stem for the P_S-type and P_I-type. The initial rubbing surface was 0.5 mm beneath the natural surface for the P_S-type and 8.5 mm for P_I-type. **Figure 1** shows the fiber orientation with the sliding direction.

Sliding wear properties of bamboo specimens were studied on a block-on-ring machine. The counterfaces were made of a gray iron (HT200) and had a diameter of 40 mm. The normal loads were from 30 to 120 N and the sliding velocity was set as 0.42 and 0.84 m s^{-1}, respectively; the total sliding distance was about 504 m and the surrounding temperature was about 23°C during all these tests. Worn morphologies of bamboo specimens and gray iron rings and wear debris were examined by SEM and stereoscopy.

2.2. Preparation of bamboo fiber specimens

Bamboos possess the excellent wear properties, and in present paper, bamboo fibers are selected as the reinforced fibers. The fresh bamboo (*Phyllostachys heterocycla*) was first cut into pieces after removing the outer and inner surface materials. The bamboo pieces were

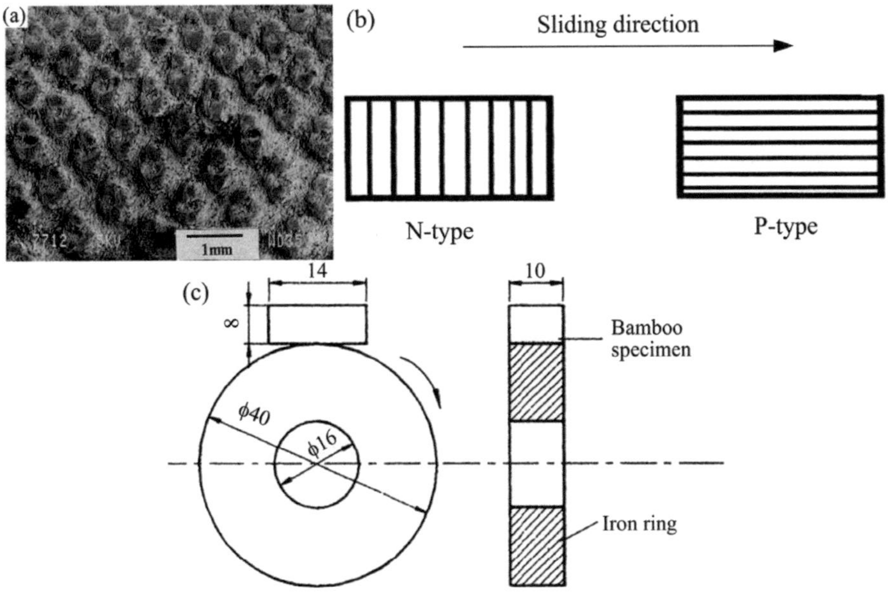

Figure 1. (a) A photograph showing section structure of bamboo stem shell; (b) schematic diagram showing fiber orientation with respect to the rubbing surface for the N and P (P_S and P_l) specimens; and (c) the contract model of block-on-ring wear.

immersed into the softener solution (2%vol $NH_3{\cdot}H_2O$, 4%vol CON_2H_4) and boiled for 1 h at 100°C. Then these pieces were kept in NaOH (4%vol) solution for 30 min at 70°C. The pieces with softened fibers were broken mechanically and carded into fiber bundle. Finally, the fibers were dried in an oven at 60°C for 30 min. Every specimens of the bamboo fiber were 40 mm (for tensile test) or 3–5 mm in length and 1–3 mm in diameter. The diameters of bamboo fibers were measured and marked using a stereo microscope, which has a precision of 0.01 μm.

2.3. Mechanical testing of the bamboo fiber

The mechanical properties of the bamboo fibers were evaluated by tensile testing machine. The specimens of the bamboo fibers for tensile tests were 40 mm in length and 0.5 ± 0.05 mm in diameter. The tensile speed was 100 mm min^{-1} referring to the pulling speed of the tensile test machine during test. Ten tests were run for each bamboo fiber. The elongation at fracture, tensile strength, and elastic modulus of the fiber were recorded.

The raw components used for preparing the friction materials are listed in **Table 1**. Phenolic resin was used as adhesive; powders (such as Al_2O_3, Sb_2S, graphite) were used as fillers. The mass fraction of the bamboo fibers added in the friction materials was 0, 3, 6, 9, and 12wt.%, respectively. All raw components were mixed using a blender for 10 min. The mixed materials were blocked with the dimension of 25 × 25 × 6 mm by compression molder equipment for 30 min at 165°C under pressure of 25 MPa, followed by heat treatment. The posttreating was segmented at 140°C for 1 h, 150°C for 3 h, and 180°C for 6 h continually as shown in **Figure 2**.

Raw ingredients	Bamboo fiber content (wt. %)				
	0	3	6	9	12
Mineral fiber	17	16.5	16.04	15.6	15.18
Glass fiber	10	9.71	9.43	9.17	8.93
Phenolic resin	13	12.62	12.26	11.93	11.61
Vermiculite powder	5	4.85	4.72	4.59	4.46
Foam iron powder	11	10.68	10.38	10.09	9.82
$BaSO_4$	20	17.47	16.98	16.51	15.57
Petroleum coke	6	5.81	5.66	5.5	5.36
Graphite	8	7.76	7.55	7.34	7.14
Al_2O_3	4	3.88	3.77	3.69	3.57
Sb_2S_3	3	2.91	2.83	2.75	2.68
Friction powder	1	0.97	0.94	0.92	0.89
Zinc stearate	2	1.94	1.89	1.83	1.79
Carbon black	3	1.9	1.1	1.08	1

Table 1. Relative contents of raw ingredients in the designed specimens of friction materials.

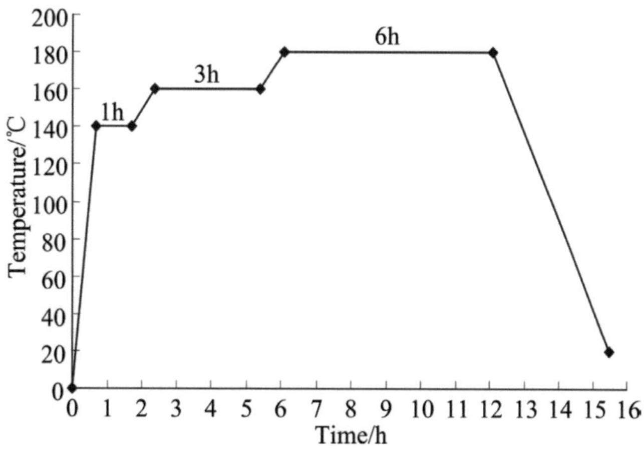

Figure 2. Heating process for preparation of friction materials.

2.4. Friction and wear tests

The tribological property of friction materials was investigated using a constant speed friction tester with speed of 7.54 m s^{-1} under pressure of 0.98 MPa. The rotating disc (HT250 cast iron) was used as the counterpart. Friction and wear tests were implemented at the test temperature of 100, 150, 200, 250, 300, and 350°C, respectively [28]. The friction coefficients ($\mu_I(i)$) (under the temperature-increasing condition) were automatically recoded. The friction coefficients ($\mu_I(i)$, $\mu_D(i)$), and specific wear rate ($V(i)$) were obtained after 5000 rotations of the disc, where $i = 1, 2, \ldots, 6$, corresponding to the temperature of 100, 150, 200, 250, 300, and 350°C, respectively. The volume wear rate of the friction materials was evaluated and calculated as follows:

$$V(t) = \frac{1}{2\pi R}\frac{A}{n}\frac{d_1 - d_2}{f_m} \tag{1}$$

where n is the number of revolutions of the disk ($n = 5000$), R is the distance between the center of the rotating disk and friction material specimen ($R = 0.15$ m), A is the contact surface between the specimen and the disk ($A = 625$ m^2), d_1 is the average thickness of specimen before test (mm), d_2 is the average thickness of specimen after test (mm), and f_m is mean value of the force.

Worn morphologies of the specimens after tests were observed using the SEM (JEOL JSM-5600) at a voltage of 25 kV.

3. Experimental results and discussion

3.1. Wear tests of bamboo specimens

Figure 3 illustrates the wear volume of the three types of bamboo specimens versus the normal load at 0.42 and 0.84 m s^{-1}. It can be seen from **Figure 3** that the wear volume was dependent upon the normal load, the sliding velocity, and the relative orientation of bamboo fibers with respect to the friction surface. The wear volume and its difference between different types of specimens increased with the increase of normal load. The wear rate at 0.42 m s^{-1} velocity was lower than that at 0.84 m s^{-1}. The N-type specimens presented excellent wear resistance. Although they had the same relative orientation of friction surface, P_S-type specimens presented better wear resistance than P_I-type specimens under identical experimental conditions, suggesting that the wear resistance of the outside layer of bamboo stem was higher than that of its inner layer.

3.2. Wear track morphologies of gray iron rings

The materials transfer from the bamboo specimens to the iron surface occurred due to adhesion. In the initial stage, transferred material formed some patches on the gray iron ring surface, as shown in **Figure 4a**. As the sliding distance was increased, transferred material patches were extended along the sliding direction because of crushing action and further adhesion. When the interfacial contact reached a steady state, the adhesion transfer film was in a relatively steady state as shown in **Figure 4b**. The transferred material film did not cover the entire friction surface of the iron ring, as the material transferred to the ring surfaces could be detached. This transferring-detaching process resulted in the adhesive wear of bamboo. Features of adhesive wear were also found on worn surfaces of bamboo specimens and will be discussed in the following sections.

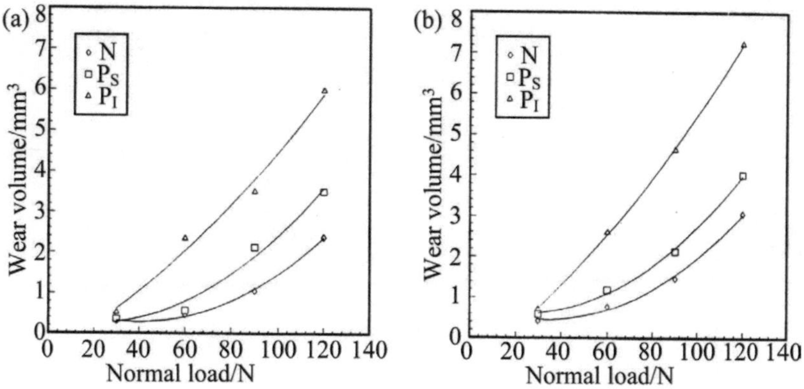

Figure 3. The wear volume of the three types of bamboo specimens versus the normal load at sliding velocity of (a) 0.42 m s^{-1} and (b) 0.84 m s^{-1}.

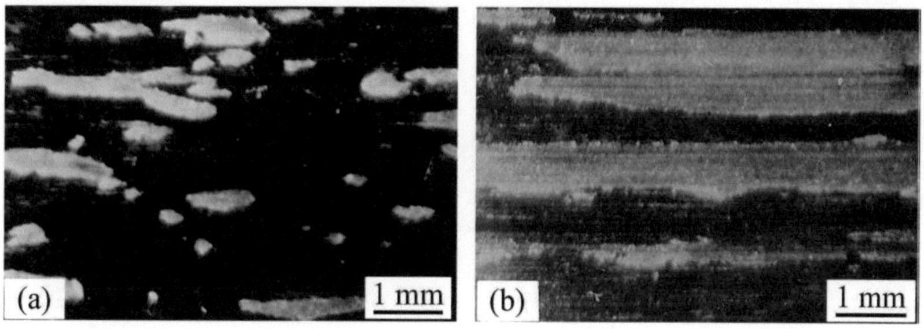

Figure 4. Stereographs of wear track of the gray iron ring.

3.3. Wear of P_S-type bamboo specimens

Figure 5 illustrates typical morphologies of worn surfaces of P_S-type specimens. There were mainly three wear features: pits, microcracks, and grooves. Pits were produced because of adhesion between bamboo specimens and iron. Some materials from these pits were transferred onto the gray iron ring surface and the remainder became wear debris. Because of asperities of the gray iron ring surface, a tensile stress existed in the bamboo surface layer at the rear of the contacting asperity and a compressive stress at front. This stress distribution could easily lead to microcracking of the bamboo surface layer across the friction direction. Moreover, the adhesion force existing at the contacting interface strengthened this microcracking process. However, microcracks along the friction direction may be directly nucleated and propagated due to the tensile stress. As the normal load was raised, the influence of microploughing-microcutting on the wear of the bamboo specimens surface layer became stronger, and the microploughing-microcutting grooves on worn surfaces generated under 60–90 N load at 0.42 m s^{-1} velocity were shallow (**Figure 5a–d**). For this case, damage of cellulose fibers was not severe. However, when the normal load reached 120 N, particularly at 0.84 m s^{-1} velocity, the cellulose fiber walls had been cut, but leptodermous

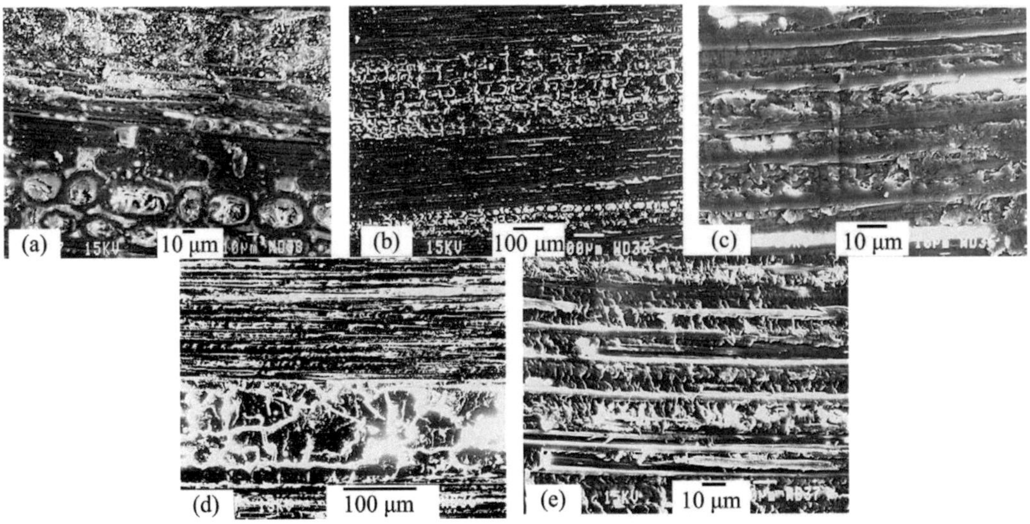

Figure 5. SEM micrographs of worn surfaces of P_S-type bamboo specimens. (a) 60 N, 0.42 m s^{-1}; (b) 90 N, 0.42 m s^{-1}; (c) details from (b); (d) 120 N, 0.42 m s^{-1}; and (e) 120 N, 0.84 m s^{-1}.

cell tissue between the cellulose fibers and the material inside vascular bundles were not completely removed (**Figure 5e**).

It can be considered from the surface morphology shown in **Figure 5** that adhesion, microcracking, and microploughing-microcutting were the main wear mechanisms of P_S-type bamboo specimens. When the normal load and sliding velocity were low, the adhesive wear, microcracking, and microploughing were predominant, while, when the load and velocity were high, the predominant wear mechanism was microploughing-microcutting.

3.4. Wear of P_I-type bamboo specimens

Figure 6 illustrates typical morphologies of worn surfaces of P_I-type bamboo specimens. It was seen that microcracks and microploughing grooves existed on the worn surfaces at low magnification, as shown in **Figure 6a** and **e**. **Figure 6c** shows some extent of the adhesive wear. At high magnification, besides some microcracks with random distribution illustrated in **Figure 6a**, some regular microcracking took place, as shown in **Figure 6c** and **e**. The regular microcracking under low load mainly displayed two states. One was local microcracking, causing some strips of bamboo material to be dug out from the surface layer, as shown in **Figure 6b**. Another was microcracking along the fiber direction (i.e., the rubbing direction) and then across fiber direction, as shown in **Figure 6e**. The tensile and compressive stresses of the bamboo surface layer under the contacting asperities of the ring surface and the adhesion force between both rubbing surfaces played important roles in the surface microcracking during the rubbing process. Generally, the former regular-microcracking occurred at the lower velocity (0.42 m s^{-1}) and the latter took place at the higher velocity (0.84 m s^{-1}). Compared with P_S-type, P_I-type specimens were severely ploughed, as shown in **Figure 6b** and **d**. When the load was increased to 120 N, particularly at high velocity (0.84 m s^{-1}),

http://dx.doi.org/10.5772/intechopen.69893

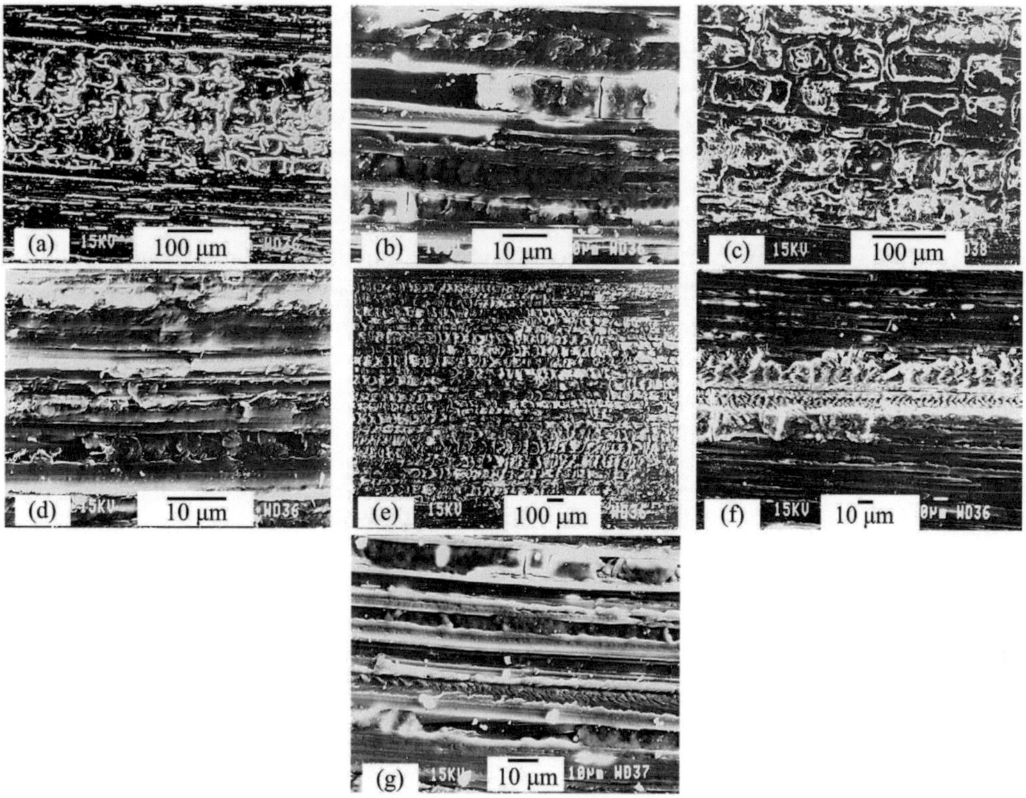

Figure 6. Worn surface micrographs of P_I-type bamboo specimens. (a) 60 N, 0.42 m s^{-1}; (b) details of ploughing grooves in (a); (c) 60 N, 0.42 m s^{-1}; (d) 90 N, 0.42 m s^{-1}; (e) 60 N, 0.84 m s^{-1}; (f) and (g) 120 N, 0.84 m s^{-1}.

asperities of the ring surface cut into the P_I-type specimen surface layer and ploughed up the middle texture material of vascular bundles (see **Figure 6f**), or very deep grooves were produced (see **Figure 6g**). These topographies represented severe wear rupture of P_I-type specimens. Comparing the morphologies illustrated in **Figures 5** and **6**, it was found that the microploughing-microcutting damage of P_I-type specimens was larger than that of P_S-type specimens under identical experimental conditions. The main wear mechanisms of P_I-type bamboo specimens were also adhesion, specially, microcracking and microploughing–microcutting. The degree of damage of P_I-type specimens was severe in comparison to that of P_S-type ones.

3.5. Wear of N-type bamboo specimens

Figure 7 illustrates typical morphologies of worn surfaces of the N-type bamboo specimens. The structure of the bamboo stem was basically revealed at low magnification, as shown in **Figure 7a, d, g,** and **h**. The larger dark zones are the ends of vascular bundles consisting of several vascular, and the remainder is matrix tissue. It can be seen that there existed a certain difference in wear behavior between vascular bundles consisting of sclerenchyma cells and matrix tissue consisting of leptodermous cells. The wear rupture of vascular bundles was severe as compared with that of the matrix tissue. This result was opposite to the abrasive wear behavior of N-type specimens in free-abrasive wear conditions, in which the ends of vascular bundles were protruded on the matrix of the abrading surface of N-type specimens, suggesting that vascular bundles of bamboo possessed higher wear resistance than matrix tissue [5]. It was seen to the naked eye that the ends of vascular bundles were darker than matrix tissue after sliding friction, particularly at high velocity or under high load (see **Figure 7d**, **e**, **g**, and **h**). High temperature caused by frictional heat would burn the contacting bamboo surface. The vascular bundles were mainly burned for the N-type. This phenomenon can be observed clearly after grinding a bamboo block (N-type mode) against an abrasive wheel. The burnt material may easily be removed by asperities of the ring surfaces. The interfacial temperature was dependent upon the normal load and, particularly, sliding velocity [29]. Therefore, topographies of the ends of vascular bundles of worn surfaces of N-type bamboo specimens varied with the load and velocity.

Figure 7b shows that the part surrounded by vascular of cellulose fibers (i.e., the center area of a vascular bundle) protruded. This was more obvious at low load and velocity (see **Figure 7a**

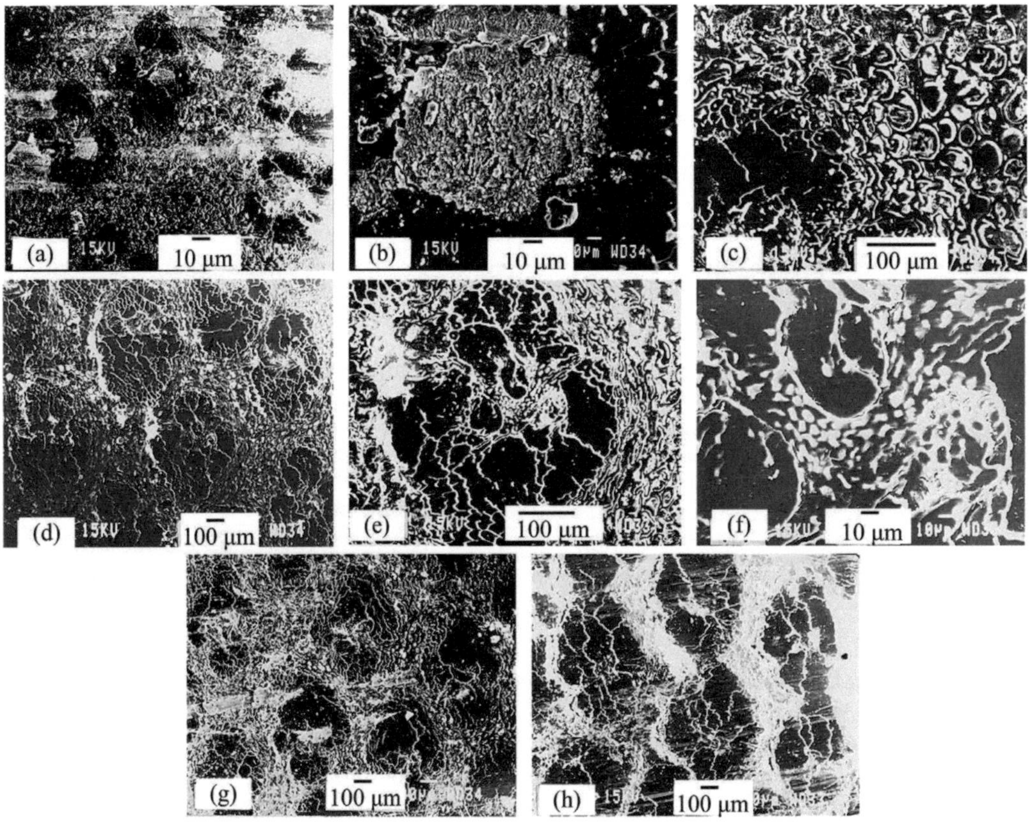

Figure 7. Worn surface micrographs of N-type bamboo specimens. (a) 60 N, 0.42 m s^{-1}; (b) details of the center area of a vascular bundle in (a); (c) details of leptodermous cell of the lower area of (a); (d) 120 N, 0.42 m s^{-1}; (e) details of a vascular bundle of the center area of (d); (f) details of the center area of the vascular bundle of (e); (g) 60 N, 0.84 m s^{-1}; (h) 120 N, 0.84 m s^{-1} (the friction direction of the ring was from left to right).

and **b**). As the load or velocity was increased, the area of the protruded part became smaller, comparing **Figure 7a**, **d**, **g**, and **h**. It was considered that the center part of a vascular bundle would have leptodermous cell texture because its tribological behavior was similar to that of matrix tissue. **Figure 7c** gives the details of the worn surface of matrix tissue.

Worn micrograph of N-type bamboo specimens mainly presented adhesive wear and microcracking. No microploughing-microcutting feature was observed on worn surfaces except those of 120 N load and 0.84 m s^{-1} velocity (**Figure 7h**). Besides the microcracking of matrix tissue, cracking between bundles and matrix occurred, as shown in **Figure 7f** and **h**.

3.6. Mechanical and physical properties of bamboo fibers

The mechanical and physical properties of bamboo fibers in this study are listed in **Table 2**. From **Table 2**, it can be seen that the elongation at fracture of alkaline-treated bamboo fibers was about 1.2%, the tensile strength was about 56.3 MPa, and the elastic modulus was about 18.6 GPa. However, for untreated bamboo fibers, the elongation at fracture was 1.7%, the tensile strength was 46.7 MPa, and the elastic modulus was 23.3 GPa. It indicated that the tensile strength and elastic modulus were increased by the modification of bamboo fibers. It is because the crystallinity of the alkaline-treated bamboo fiber was increased, which could improve the polarity of molecules and the adhesion strength between the high molecules. The slip produced by destruction of the binding force of molecules was relatively small, so the elongation at fracture was reduced, and the tensile strength and elastic modulus was increased after modification for bamboo fiber.

The stress-strain curve of the alkaline-treated bamboo fibers is shown in **Figure 8**. At the initial stage, the stress was proportional to the strain, which is consistent with the Hooke's Force Law. There was no yield and necking phenomenon, and the stress and strain were very low before breaking. It indicated that the bamboo fiber was a brittle material. The brake pads were basically acted by compression pressure when they work. Hence, the lower tensile strength of bamboo fiber does not have impact on the braking performance of brake pad.

It can be seen from **Figure 9a** that a major part of the fracture surfaces presented brittle failure. Some small molecule impurities on bamboo fiber surface were removed by the alkaline solution that caused the adhesion strength among fibers to be reduced considerably. The fracture of the untreated fiber (**Figure 9b**) showed that the uneven break presented because of the uneven stress that resulted from the bonding of pectin with lignin among the fibers.

The bamboo fibers connected with each other under pressure. The tangential resistance generated in relative sliding is called friction force. The tangential resistance is called cohesive force at the normal press of zero. **Figure 10** shows that the bamboo fibers were assembled, entangled, bonded,

	Elongation (%)	Tensile strength (MPa)	Elastic modulus (GPa)
Untreated bamboo fiber	1.7	46.7	18.6
Alkaline-treated bamboo fiber	1.2	56.3	23.3

Table 2. Test results of mechanical properties of bamboo fiber.

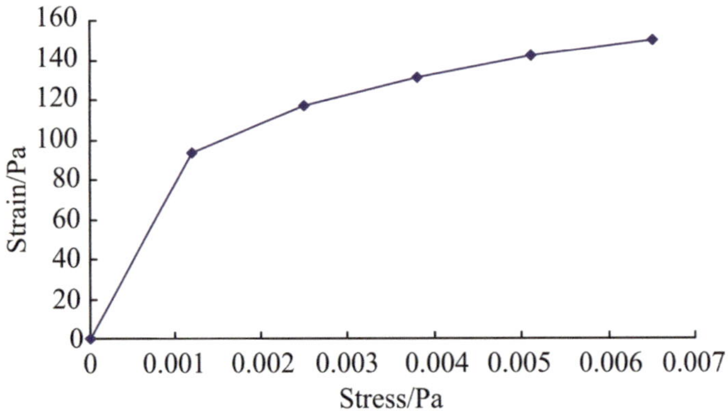

Figure 8. Test result of tensile stress-strain curve of a bamboo fiber treated with alkaline solution.

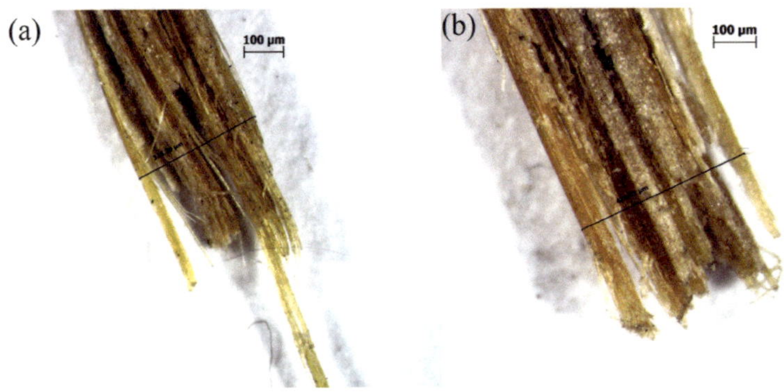

Figure 9. Morphologies of tensile fracture of (a) the bamboo fiber treated with alkaline solution; and (b) the untreated bamboo fiber.

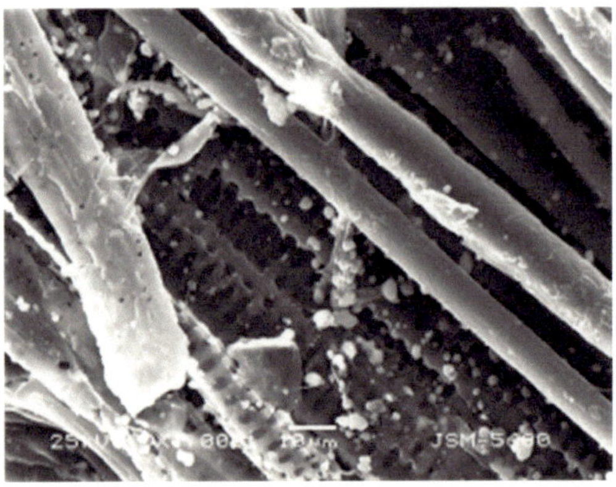

Figure 10. Bamboo fiber assemblies.

and held tightly together due to the role of cohesive force that is not easy to loose. Therefore, the properties of the friction materials were affected by the friction force and the cohesive force.

3.7. Effect of bamboo fiber on friction performance

It can be found from **Figure 11** that the friction coefficients of BFRFMs with 3, 6, and 9 wt.% bamboo fibers were higher than those of the friction materials without bamboo fibers. The friction coefficients of the BFRFMs have significant variations at the test temperature of 250°C. This is because phenolic resin began to pyrolyse, and the bamboo fibers were carbonized gradually when temperature exceeded 250°C. However, the friction coefficient of BFRFMs containing 12 wt.% bamboo fibers decreased with the increase of test temperature.

It can be seen from **Figure 12** that wear rates of the BFRFMs generally increased with the increase of test temperature since the matrix began to soften, and the bamboo fibers were carbonized with the increase of test temperature. The surface roughness of friction materials containing 6, 9, and 12 wt.% bamboo fibers was high, so the adhesive wear and microcutting wear appeared on the worn surface. This is because the heat fading of the phenolic resin appeared, and the small hard particles formed from some glass fibers separated from the matrix. Plenty of wear debris formed and fell off, so the wear rate of friction materials significantly increased. The wear rate of friction materials containing 3 wt.% bamboo fibers was the lowest. In fact, some voids and grooves formed after the carbonization of the bamboo fibers can contain some other abrasive particles.

3.8. Wear surface morphologies of BFRFMs

The worn surface morphologies of the BFRFMs are shown in **Figure 13**. It can be seen from **Figure 13** that some glass fibers exposed and some did not separate from the matrix. The glass fibers and friction surfaces were supported by the matrix. Some hard particles from the glass

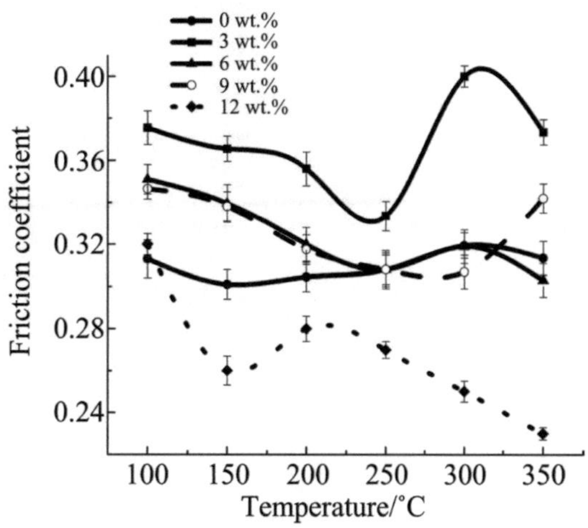

Figure 11. Variation of the friction coefficient of the bamboo fiber reinforced friction materials with the temperature.

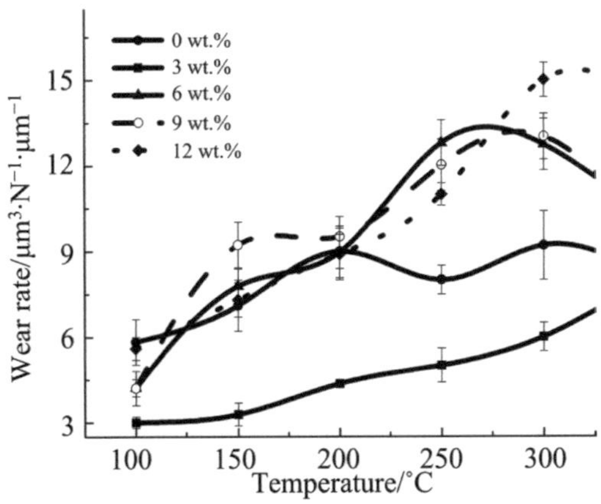

Figure 12. Variation of wear rate of the bamboo fiber reinforced friction materials with temperature.

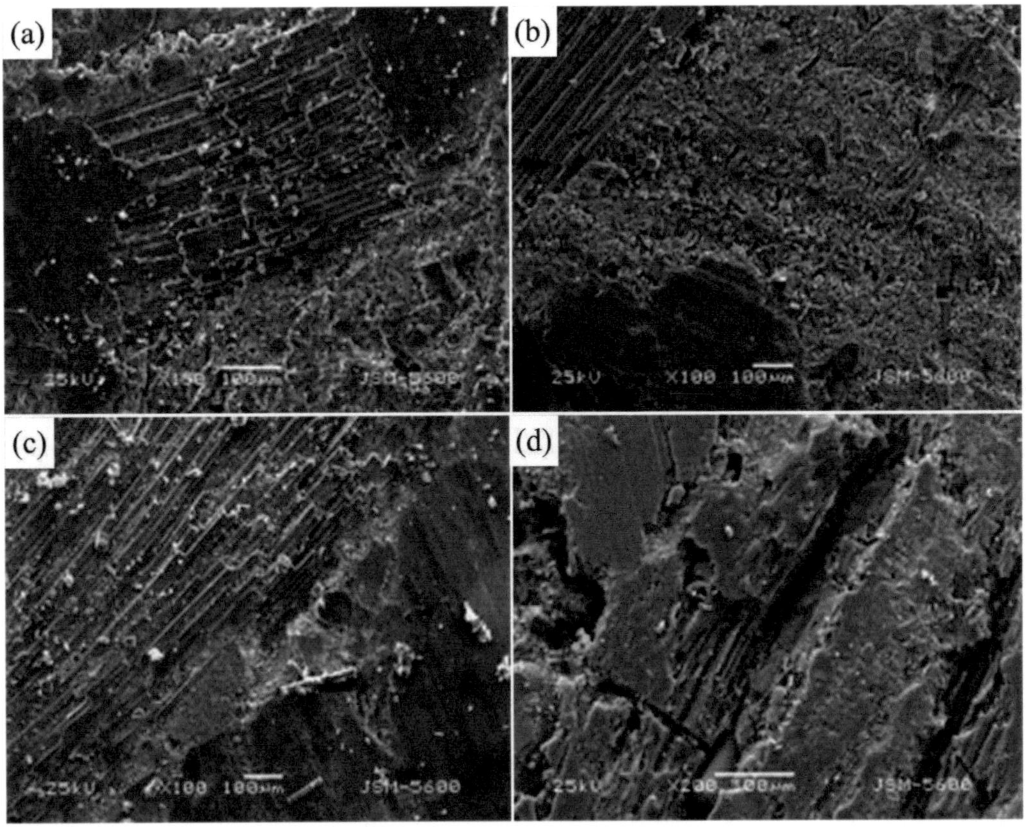

Figure 13. Surface morphologies of the BFRFMs with bamboo fibers of (a) 3 wt.%; (b) 6 wt.%; (c) 9 wt.%; and (d) 12 wt.%.

Figure 14. Morphologies of the groove and voids of the BFRFMs with bamboo fibers of (a) 3 wt.%; and (b) 6 wt.%.

fiber formed under the friction force and the glass fibers did not separate from the matrix completely. It illustrated that the glass fibers were firmly bound with the matrix. Graphite and some particles that carbonized existed on the friction surface, so part of the friction surface was very smooth and easily deformed under friction force and shear force. Therefore, wear resistant surface was formed, and the friction coefficient and wear rate were decreased. Meanwhile, the carbonized fiber can repair the scratch and shallow pits on the worn surface, so the adhesive wear was decreased to some extent [30, 31], and the worn surface is relatively smooth. The grooves or voids (**Figure 14**) formed after the carbonization of the fibers reduced the noise and adhere to wear debris on the wear surface of friction materials [32].

4. Conclusion

a. Dry sliding wear of bamboo (*P. pubescens*) stem is dependent upon the normal load, the sliding velocity, and the cellulose fiber (vascular bundle), and orientation with respect to the rubbing surface. The wear volume of bamboo increases with the crease of the normal load and sliding velocity. Normal-oriented specimens exhibit better wear resistance than parallel-oriented ones, and the outside surface layer has better wear resistance than the inner layer.

b. Material transfer phenomena from bamboo to the counterface occur for three types of bamboo specimens. The predominant wear mechanisms are adhesion, microcracking, and microploughing under low load at low velocity, and microploughing-microcutting under high load for P_S-type specimens. The main wear mechanisms of P_I-type specimens are also adhesion, and particularly microcracking and microploughing-microcutting. The wear of N-type specimens is mainly due to adhesion and microcracking, and the matrix tissue shows certain wear resistance.

c. The friction coefficients of friction materials containing 3, 6, and 9 wt.% bamboo fibers increased with increase of the test temperature, whereas the friction materials containing 12 wt.% bamboo fibers decreased.

d. The wear rates were decreased with the increase of bamboo fibers content when the content of bamboo fibers of friction materials was lower than 3 wt.%; the wear rate increased with increase of bamboo fibers content when the bamboo fibers content of friction materials was higher than 3 wt.%.

e. The bamboo fibers improved the friction performances of fiction materials. The grooves or voids formed after the carbonization of the fibers reduced the noise and adhere to wear debris on the wear surface of friction materials.

Acknowledgements

This study was supported by National Natural Science Foundation of China (Grant No. 51475205), by Jilin Province Science and Technology Development Plan Item (Grant No. 20150519022JH), by China-EU H2020 FabSurfWAR project (Grant Nos. S2016G4501 and 644971), by Special Foundation of National Key R&D Program of China (2016YFD0701601), and by Supporting Program of "the twelfth five-year-plan" in national science and technology in rural areas of China (Grant No. 2014BAD06B03-01). Parts of this chapter are reproduced from authors' recent conference publication, work *etc.* [29, 32].

Author details

Yunhai Ma[1,2], Yucheng Liu[1,3] and Jin Tong[1,2]*

*Address all correspondence to: jtong@jlu.edu.cn

1 Key Laboratory of Bionic Engineering (Ministry of Education, China), Jilin University (Nanling Campus), Changchun, China

2 College of Biological and Agricultural Engineering, Jilin University (Nanling Campus), Changchun, China

3 Editorial Department of Journal of Bionic Engineering, Jilin University (Nanling Campus), Changchun, China

References

[1] Lakkad C, Patel JM. Mechanical properties of bamboo, a natural composite. Fibre Science and Technology. 1981;**14**:319-322

[2] Godbole V S, Lakkad SC. Effect of water absorption on the mechanical properties of bamboo. Journal of Materials Science Letters. 1986;**5**:303-304

[3] Li S, Zhou B, Zheng Z, Zeng Q. A fine-scale bionic model for composite materials. Chinese Journal of Materials Research. 1991;**5**:542-547

[4] Yakou T, Sakamoto S. Abrasive properties of bamboo. Japanese Society of Tribologists. 1993;**38**:491-497

[5] Tong J, Ren LQ, Li JQ, Chen BC. Abrasive wear behaviour of bamboo. Tribology International. 1995;**28**:323-327

[6] Bijwe J. Composites as friction materials: Recent development in non-asbestos fiber reinforced friction materials—a review. Polymer Composites. 1997;**18**:378-396

[7] Xiao G, Zhu Z. Brake materials development by using DOE/RSM and artificial neural network. Tribology International. 2010;**43**:218-227

[8] Hong US, Jung SL, Cho KH, Cho MH, Kim SJ, Jang H. Wear mechanism of multiphase brake materials with different phenolic resin matrices. Wear. 2009;**266**:739-744

[9] Almaslow A, Ratnam CT, Ghazali MJ, Talib RJ, Azhari CH. Effects of electron-beam and sulfur crosslinking of epoxidized natural rubber on the friction performance of semimetallic friction materials. Composites Part B: Engineering. 2013;**54**:377-382

[10] Cho MH, Ju J, Kim SJ, Jang H. Tribological properties of solid lubricants (graphite, Sb_2S_3, MoS_2) for automotive brake materials. Wear. 2006;**260**:855-860

[11] Kukutschová J, Roubíček V, Mašláň M, Jančík D, Slovák V, Malachová K, et al. Wear performance and wear debris of semimetallic automotive brake materials. Wear. 2010;**268**:86-93

[12] Chang QX, Wang KS, Zhao HJ. The friction and wear properties of short Carbon/CNT/PA6 hybrid composites under dry sliding conditions. International Journal of Polymeric Materials. 2013;**62**:540-543

[13] Yousif BF, Nirmal U, Wong KJ. Three-body abrasion on wear and frictional performance of treated betel ≥nut fibre reinforced epoxy (T-BFRE) composite. Materials & Design. 2010;**31**:4514-4521

[14] Hashmi SAR, Dwivedi UK, Chand N. Graphite modified cotton fibre reinforced polyester composites under sliding wear conditions. Wear. 2007;**262**:1426-1432

[15] Eleiche AM, Amin GM. The effect of unidirectional cotton fiber reinforcement on the friction and wear characteristics of polyester. Wear. 1986;**112**:67-78

[16] Ma YH, Liu YC, Ma SS, Wang HB, Shang WB, Gao ZH, et al. Friction and wear properties of dumbbell-shaped jute fiber-reinforced friction materials. Journal of Applied Polymer Science. 2014;**131**. DOI: 10.1002/app.40748

[17] Liu L, Wang Q, Xia Z, Yu J, Cheng L. Mechanical modification of degummed jute fibre for high value textile end uses. Industrial Crops and Products. 2010;**31**:43-47

[18] Chand N, Dwivedi UK. Effect of coupling agent on abrasive wear behaviour of chopped jute fibre-reinforced polypropylene composites. Wear. 2006;**261**:1057-1063

[19] Chen Y, Sun LF, Ovidiu C, Ioan N, Yachmenev V, Mary W. Kenaf/ramie composite for automotive headliner. Journal of Polymers and the Environment. 2005;**13**:107-114

[20] Wu CR, Jia SK, Chen RY, Huang Z, Zhai SF, Feng YH, et al. Composites of sisal fiber/polypropylene based on novel vane extruder: Effect of interface and damage on mechanical properties. Journal of Reinforced Plastics and Composites. 2013;**32**:1907-1915

[21] Mohan TP, Kanny K. Water barrier properties of nanoclay filled sisal fiber reinforced epoxy composites. Composites: Part A. 2011;**42**:385-393

[22] Liu Q, Hughes M. The fracture behavior and toughness of woven flax fiber reinforced epoxy composites. Composites: Part A. 2008;**39**:1644-1652

[23] El-Tayeb NSM. A study on the potential of sugarcane fibers/polyester composite for tribological applications. Wear. 2008;**265**:223-235

[24] Paramesaran N, Liese W. On the fine structure of bamboo fibers. Wood Science and Technology. 1976;**10**:231-246

[25] Fuentes CA, Tran LQN, Dupont-Gillain C, Vanderlinden W, De Feyter S, Van Vuure AW, et al. Wetting behaviour and surface properties of technical bamboo fibers. Colloids and Surfaces A: Physicochemical and Engineering Aspects. 2011;**380**:89-99

[26] Dieu T, Liem N, Mai T, Tung N. Study on fabrication of BMC laminates based on unsaturated polyester resin reinforced by hybrid bamboo/glass fibers. JSME International Journal Series A. 2004;**47**:570-573

[27] Nahar S, Khan RA, Dey K, Sarker B, Das AK, Ghoshal S. Comparative studies of mechanical and interfacial properties between jute and bamboo fiber-reinforced polypropylene-based composites. Journal of Thermoplastic Composite Materials. 2012;**25**:15-32

[28] Ma YH, Liu YC, Menon C, Tong J. Evaluation of wear resistance of friction materials prepared by granulation. ACS Applied Materials & Interfaces. 2015;7:22814-22820

[29] Tong J, Arnell RD, Ren LQ. Dry sliding wear behaviour of bamboo. Wear. 1998;**221**:37-46

[30] Essabir H, Achaby MEI, Hilali EIM, Bouhfid R, Qaiss A. Morphological, structural, thermal and tensile properties of high-density polyethylene composites reinforced with treated Argan nut shell particles. Journal of Bionic Engineering. 2015;**12**:129-141

[31] Wen S, Li K, Song Q, Shan Y, Li Y, Li H, Ma H. Enhancement of the oxidation resistance of C/C composites by depositing SiC nanowires onto carbon fibers by electrophoretic deposition. Journal of Alloys and Compounds. 2015;**618**:336-342

[32] Ma YH, Shen SL, Tong J, Ye W, Yang YZ, Zhou J. Effects of bamboo fibers on friction performance of friction materials. Journal of Thermoplastic Composite Materials. 2013;**26**:845-859

Chapter 7

The Use of Bamboo for Erosion Control and Slope Stabilization: Soil Bioengineering Works

Guillermo Tardio, Slobodan B. Mickovski,
Hans Peter Rauch, Joao Paulo Fernandes and
Madhu Sudan Acharya

Additional information is available at the end of the chapter

http://dx.doi.org/10.5772/intechopen.75626

Abstract

The potential of bamboo in erosion control and slope stabilization has been proven worldwide. Bamboos are being used as living plants as well as construction material in different soil bioengineering techniques in many countries. The soil and water bioengineering approach is combined with bamboo traits and mechanical properties. The existing accumulated experiences of using bamboo in soil and water bioengineering works, along with the existing standards and design guidelines, make bamboo species an essential and cost-effective material for erosion control and slope stabilization works. In this chapter, all the necessary aspects to be taken into account for an appropriate use of bamboo in soil bioengineering works are addressed, and the design approaches for soil bioengineering works using bamboos are presented.

Keywords: bamboo, soil bioengineering, erosion control, slope stability, soil strength

1. Introduction

Bamboo is a globally distributed group of plants with more than 1400 species distributed worldwide in tropical, equatorial and semitropical biomes.

It builds important and diversified habitats with different specificities, according to the nature of the species and the general ecological conditions.

Most bamboo species show a very strong development and colonization ability, determining that in some temperate habitats, they can assume an invasive character.

The nature and characteristics of some bamboo communities can present an important ability to soil and slope protection and stabilization, as one can easily confirm by observing developed bamboo forests in mountainous areas with very steep slopes.

On the other hand, the structural characteristics offered by some woody bamboo species make bamboo a valuable basic construction material in many regions (e.g. India, China and Southeast Asia).

These characteristics determine that these species and communities can be of high interest for soil and slope protection and reinforcement works, particularly in areas where the bamboo is native. They can be used integrated and fostering natural communities, ensuring efficient soil cover and reinforcement functions through their high, lightly dense culms and their dense and resilient root systems.

These functions are also of particular interest for soil bioengineering because bamboo has biological characteristics such as a high vegetative propagation ability (making its reproduction very easy) and a rapid growth (allowing for a quick effect on soil cover and root consolidation). Moreover, the structural and physical characteristics of the stems of certain bamboo species turn them into a very effective construction material for complementary soil bioengineering support structures.

Therefore, bamboo in its different forms and associated communities, and also in terms of particular species, is of high interest for nature and biodiversity protection (and therefore ecosystem restoration) as well as for slope protection and stabilization.

2. Principles of soil bioengineering

Soil bioengineering comprises a diversified group of techniques and land management systems developed by mankind throughout the millennia to use natural systems and elements in order to ensure the safety and functionality of land uses in a context of restricted availability of materials and, particularly, energy.

Soil bioengineering techniques have been used throughout the world with the available plants and construction systems, many times replicated in different continents due to its efficiency and easy construction.

Only in the first decades of the twentieth century, this set of building and land management techniques has been recognized as an integrated engineering approach to many soil stabilization problems, and they started to be systematized, studied and developed.

This situation led to the development of an engineering discipline where 'soil bioengineering has set itself the aim of designing our environment in a "living" way by applying construction methods which are close to nature (…) based on materials which are found in nature and which are combined with technical building materials' [1].

This engineering domain developed from the rediscovery of traditional building and management techniques that use predominantly living plants and vegetation communities as building materials, nowadays, presents a strong evolution with the development of new materials and plant/material combinations, building techniques and innovative domains of application. This domain of engineering combines classical areas of civil engineering (e.g. structures, materials, construction, geological engineering) with biology, integrating a wide diversity of disciplines and specialization domains. The aim is to achieve feasible, efficient durable, self-repairing, resilient, evolving and ecological functional engineering structures that, within strict technical and geotechnical limits, normally fulfil their planed functions with higher efficiency and lower cost [2].

The European Federation for Soil Bioengineering (EFIB) defines soil and water bioengineering as:

> *'Typically, plants and parts of plants are used as living building materials, in such a way that, through their development in combination with soil and rock, they ensure a significant contribution to the long-term protection against all forms of erosion. In the initial phase, they often have to be combined with non-living building materials, which may, in some cases, ensure more or less temporarily, most of the supporting functions.*
>
> *The use of organic materials is preferred, because parallel to the development of the vegetation and its increasing stabilization ability, these materials will rot and be reincorporated in the natural biogeochemical cycles. Also preferred are indigenous (autochthonous) and site-specific plants, as they promote a biodiversity suited to the landscape. The planning and construction objectives are the protection and stabilization of land uses and infrastructures as well as the development of landscape elements'* [3].

Soil bioengineering aims, therefore, at ensuring an efficient nature-based solution to the protection of infrastructures. This can be helpful in situations of conflict between opposite needs: the human demand for larger spaces for activities and infrastructures and the natural systems intrinsic need for development space.

Soil bioengineering systems use plants and parts of plants as living building materials as well as introducing and developing functional living communities that are able (*per se* or complemented by a wide variety of materials and structures) to effectively ensure the desired soil and slope protection and consolidation targets. Its goal, as referred therewith, is mainly functional in terms of protecting and integrating within the infrastructure, as well as landscape protection and restoration. Its particular characteristic is the fact that the result of its application is not an inert structure, but a dynamic, resilient living community able to restore itself after disturbances and, if adequately maintained, to ensure a long-term, non-decaying, effective intervention with permanently developing efficiency.

Due to the nature, characteristics and properties of vegetation, it is important to note that bioengineering strategies also have limitations in terms of their effectiveness and application limits. The first one is that only a limited available number of plants from a given habitat have the necessary technical characteristics constraining the potential use of the aimed technical solutions. Secondly, plants, as living organisms, do not behave in a standardized way, limiting the ability to precisely calculate the technical effectiveness of the interventions. Finally, plants have limited ability in terms of root growth, hindering their capacity to stabilize soils to depths larger than 1.5–2 m, depending on the species. It is also important to note that there is a lack of a systematized knowledge on the physical behaviour of plants and particularly of their roots and root systems, when exposed to external forces, despite the promising results of an ever-growing research effort.

These limitations imply the need for the use of complementary structures to help overcome—temporarily or permanently—the local adverse conditions. This situation determined the development of a particular segment of the industry related to complementary materials (e.g. organic geotextiles) aimed at reducing the impact of water and soil erosion in the initial development phases of the construction and interventions and to the conception of construction techniques using classical civil engineering approaches and materials in combination with the advantages brought by vegetation.

The main concerns for soil bioengineering are related to soil support, cover, and consolidation, as well as the regulation of the forces and processes (mainly hydrological, hydraulic, and wind-related) that act as disturbance factors.

The main functions fulfilled by the bioengineering approach are the following:

- Support functions, in terms of building or fostering the development of structures able to stabilize slopes affected either by an increase on their slope angle or by an increase of the external or internal acting forces
- Cover functions in terms of protection against erosion and trampling
- Consolidation functions in terms of soil protection, structuring, and reinforcement
- Regulation functions in terms of hydrological processes such as interception, evapotranspiration, infiltration, and runoff control

These functions are performed mainly through the action of plant aerial parts and roots as well as the associated soil biota, through their action in soil anchoring, structuring, aggregating, draining, buttressing, and reinforcement. All of these functions, mainly ensured by living autochthonous vegetation, have the complementary advantages of promoting biodiversity and strongly reducing the CO_2 emissions, not only through its capture during construction but also because the techniques and the nature and quantities of the complementary materials used imply a lower production of greenhouse gases and natural resource consumption.

3. Use of bamboo in soil bioengineering techniques

The strength of bamboo culms and roots and their straightness, lightness combined with hardness, range and size of hollowness make them potentially suitable for a variety of both structural and nonstructural applications. With good physical and mechanical properties, low shrinkage and good average density, bamboo is well suited to replace wood/timber in soil bioengineering applications but also to act on its own as a living material providing rapid ground coverage and sediment trapping, increasing surface roughness, increasing soil strength and decreasing pore-water pressures in the soil by evapotranspiration.

The selection of appropriate techniques is based on a specific site assessment and design criteria. Local climate conditions (precipitation regimes, seasonal variation, averages and extremes of temperature and rainfall), topography (slope gradient, terrain shape, elevation, sun exposition), soil (types, permeability, moisture and nutrient conditions), hydrological conditions and the most relevant erosion processes define the set of feasible techniques for a particular site. In a following step, the evaluation of the existing surrounding vegetation is most important for the design, in terms of project limitations, opportunities and potential long-term achievements. Even when bamboo is the main vegetal constructive element, the long-term success of any bioengineering implementation work is based on a wide range of plant species. It is also important to take into account the bioengineering-specific local logistical and economic constraints. Finally, all this gathered and specific site information forms the basis for selection of the appropriate bioengineering technique, plants and materials to use.

The use of bamboo to make retaining structures for soil mass or for stream bank erosion control has been practiced in traditional way in various places around the world for long time. Live bamboo stakes, wattle fence, hedge brush layering techniques and bamboo crib walls are most commonly used bioengineering techniques. Several handbooks describe these techniques and can be used as references [4]. However, an engineering design of bamboo retaining structures, such as bamboo crib wall, has not been detailed so far. For this reason, this section is focused on this bamboo bioengineering technique.

A live bamboo crib wall is a three-dimensional structure created from untreated bamboos, fill material and live cuttings. Morgan and Rickson [5] described the crib wall as 'a specialized form of gravity-retaining structure using on-site fill material, held within a constructed framework, to provide most of the necessary mass to resist overturning by the weight of both the slope and the materials'.

This crib structure, once filled, acts as a retaining structure and supports the slope. The bamboo and other installed plants provide immediate protection and stability to the structure. However, it has to be taken into account that the structure stability and resistance to failure will be gradually decreasing as its construction materials decompose. As the bamboo elements of a crib wall decompose, the live cuttings of plants or bamboo clumps will grow and proliferate. The resulting root mass will then bind the fill material and the parent soils of the slope into a single continuum, which will have enhanced strength and contribute towards the stability of the slope. **Figure 1** shows the construction steps from a practical application of a bamboo crib wall in Nepal [6, 7].

3.1. Live bamboo crib wall materials and construction specification

Both freshly cut and seasoned bamboos can be used in the construction of crib walls. Additionally, lime-treated or chemically treated bamboo stems can also be used. However, the bamboo treatments can make the crib wall construction very expensive. Therefore, it is suggested either to use freshly cut green bamboo or air-dried bamboo for crib construction.

The twig and large knots need to be trimmed although it is not necessary to make the stem smooth. Based on the size of the bamboo stems available at the site, it can be used as single stem or a bundle of three bamboos to make the header and stretcher elements in the crib construction. If larger diameters are used, the bamboo crib wall will resemble a wooden log crib wall, and the crib construction procedure will be similar to wooden log crib wall. If single bamboo stem is used for the header and stretcher elements, it is recommended to use uniform-sized bamboo stems to ensure a uniform thickness of the crib layer.

Figure 1. Crib wall construction steps in Thankot (Nepal) 2002 [7].

To secure the bamboo stems together in a bundle or to make the crib form, suitable binding wires, binding materials or nails should be used so that the bamboo stems will not tear apart or break/bend along their length.

In addition to the bamboo and binding wires, the live cuttings or rooted plants (if possible and according to the nature of the plant, longer than the depth of the crib wall to allow penetration into the soil at the back) and suitable fill materials are also required for the construction of crib walls. Generally, locally available slope or cut material is used as fill material for the crib wall construction. However, if the material contains very coarse gravel where roots cannot develop, a layer of fine, organic-rich material or humus material should be placed around the cuttings to promote the root growth. Large stones and boulders are not recommended as fill materials. The fill material should allow some degree of compaction, so that there will be no large voids within the body of the crib structure.

Before starting the crib wall construction, bamboos, binding wires and cuttings should be stockpiled near the construction site. After setting out the layout of the wall, the foundation for the crib wall construction should be prepared according to the requirement of designed dimensions of the crib wall. The depth of foundation trench shall be between 0.3 and 0.5 m, depending upon the height of the bamboo crib wall. The foundation should be inclined at about 10–15° inwards (in-slope) to increase the stability against sliding. It is recommended to have at least one layer of cribs below the existing ground level to prevent the structure from sliding and the foundation from getting undermined by any seepage water.

3.1.1. Construction procedure

First, the header and stretcher elements should be prepared by making bundles of three bamboo stems of uniform size. The header elements are cut according to the designed length, which defines the total width of the crib wall. After the stretcher and header crib elements are prepared, the first two stretcher elements should be laid on the prepared foundation trench, parallel to each other. The spacing of these stretcher elements will be controlled by placing the header elements in a specified interval as per the design drawings and specifications. After laying the header and stretcher elements, they should be firmly bound together to make a crib frame. The first layer of the crib frame should then be secured by bamboo pegs of appropriate sizes and should be bound with a peg.

After the completion of the first layer of the frame, fill materials should be placed inside the crib frame and compacted by using hand rammer. The fill material should be compacted to about 50–70 mm above the crib elements. After placing the fill, cuttings or seedlings of rooted plants should be placed horizontally on the top of the fill material. The spacing of cuttings should vary between 100 and 300 mm depending upon the size of cuttings and species of the plants used. While placing the second crib layer, the second crib frame should be set inwards to maintain the designed front batter. In this way, a stable vegetative bamboo crib wall can be constructed. After the completion of the topmost layer, an additional layer of soil should be added to ensure flush finish with the existing slope of the ground.

3.1.2. *Maintenance*

After the crib wall construction, in dry weather and low soil moisture, it is recommended to irrigate the cuttings and other plants. After the successful establishment of the cuttings, during the initial growing stages, they should be protected from grazing, so that the young plants can grow undisturbed. If some cuttings do not thrive, additional cuttings should be inserted or planted on the wall to ensure green cover of the front face of the wall and stop the washing out of the structure's fill material. If the crib wall settles due to consolidation, an additional fill material should be added on the top of the wall. The vegetated crib wall should be well maintained for at least the first 2 years after construction (**Figure 2**).

In this bioengineering technique, bamboo and plant materials are used as structural elements to take some load and resist earth pressure. This type of crib wall could be used as an alternative to gabions or masonry retaining walls.

Figure 2. View of slide at Badikhel/Lalitpur, Nepal, in May 2003 and August 2003 [7].

4. Biotechnical properties of bamboo

The plants' biotechnical properties are those plant traits contributing to a good performance and effectiveness of the bioengineering work.

4.1. Biomass growth: aerial and root system characteristics

Bamboo belongs to the grass family and has an aerial part characterized by a jointed stem called a culm. The culms are typically hollow with the exception of certain bamboo species which have solid culms. Each culm segment begins and ends with a solid joint called a node. It is in these nodes that the vegetative parts of the culm able to develop the culm vertically, produce branches and develop roots and stems if stacked or laying in the ground can be found. The underground part of the plant is built from rhizomes growing normally at a shallow depth (up to a maximum of 150 mm) from where the roots develop. These roots can grow deep into the soil up to 500 mm. The rhizomes are the main form of spreading of the plant by growing horizontally away from the plant and, because they have a similar structure as the culm with vegetative nodes developing either roots or buds, originate new shoots and new individuals.

Bamboo is the fastest growing perennial, evergreen, arborescent plant with a resulting high productivity: the dry weight yield per hectare could total as much as 32–38 or even 47 tons of biomass per hectare per year but averaging 8–18 tons per ha per year in normal conditions according to the different species and locations [8].

This productivity, expressed both for the aerial and the root parts of the plant, illustrates the ability of bamboo to cover the terrain very rapidly, to develop a dense network of subsuperficial rhizome and root system which would structure and consolidate the upper soil layer. The growth rate of each plant varies, but there are references of a 900 mm culm elongation in 1 day. The growth rate (both of the culm, the rhizome and root system, buds and shots) corresponds to a vegetative cycle that varies with the species and the climatic conditions. The growth factors (like starch reserves on the culm and the rhizome) vary with the evolution of the growth season but are maximal before sprouting, meaning a high resilience to disturbance and regeneration ability. The biomass production is very intensive both above and below ground with values of above ground dry weight varying between 0.8 and 1 tons per ha in some references [9].

The bamboo stands act as an important factor in water and nutrient conservation, as well as soil protection and runoff control. There are references of reduction in nutrient loss higher than 50% and similar values for runoff retention. This shows that bamboo stands, although having little geotechnical ability in terms of slope stabilization (due to the low depth of rooting), could have a very important role in local water cycle regulation and, therefore, soil consolidation and stabilization, preventing erosion and reducing infiltration [10].

4.2. Bamboo natural distribution

The use of native plant species is inherent to the bioengineering work approach and philosophy. Moreover, the use of indigenous species is a compulsory feature of the living material used in these works. Hence, the knowledge of bamboo natural distribution is necessary for its use.

Published literature notes approx. 1400 different species (grouped in one herbaceous and three woody types) identified and designated as bamboo (https://www.eeob.iastate.edu/research/bamboo/index.html).

Bamboo is globally distributed between 51°N and 47°S, particularly in subtropical, tropical and equatorial regions. It also covers a high altitude range, reaching up to 4000 m above sea level and thriving at temperatures as low as −20°C. The main area of occurrence is Asia where the largest number of species can be found.

This wide distribution does not mean that all species or even natural stands thrive without problems. Many forest stands are being intensively exploited, endangering a high number of species, namely, several mammal, bird and even bamboo species all classified as 'endangered' by IUCN. Several hundred species of bamboo occupy remaining natural forest stands not bigger than 2500 km^2 [11].

Due to its versatility, physical characteristics, rapid growth, and easy establishment, it is intensively exploited, not only in terms of harvesting natural forests but also in growing areas of cultivation, where there is a selection of the economically more attractive species (few dozens). The exploitation is mainly located in China, India and Southeast Asia (but also, with increasing importance, in Central and South America) and predominantly aimed at species with applicability in construction or other industries (e.g. paper pulp or laminated and other composite productions, biomass production) [12].

There is also a growing interest for bamboo as an ornamental plant, which brought the spread of several species to areas outside their natural ecological areas. This also raised some problems such as turning into invasive species and threatening natural habitats [13].

4.3. Mechanical properties of bamboo for soil bioengineering applications

The strength of bamboo culms, their straightness and their lightness combined with hardness, range and size of hollowness make them potentially suitable for a variety of both structural and nonstructural applications. With good physical and mechanical properties, low shrinkage and good average density, bamboo is well suited to replace wood/timber in soil bioengineering applications.

For example, from more than 100 bamboo species native to India, only around 20 have been systematically tested, and 16 have been found to be adequate for use in construction (**Table 1**, [14]).

The compressive strength of bamboo ranges between 35 and 70 N/mm^2 which is twice to four times the value of most timber species. The range can be explained by the different test methods and used samples. Bamboo with low moisture content has a higher compressive strength than bamboo with high moisture content (https://www.bambooimport.com/en/blog/what-are-the-mechanical-properties-of-bamboo).

The average tensile strength of bamboo is approximately 160 N/mm^2 which is around three times higher than most conventional construction grade timber materials (https://www.bambooimport.com/en/blog/what-are-the-mechanical-properties-of-bamboo).

Species	Properties							
	In green condition				In air dry condition			
		Modulus of rupture	Modulus of elasticity	Max. compressive strength	Density	Modulus of rupture	Modulus of elasticity	Max. compressive strength
	kg/m$^{3kg/m3}$	N/mm$^{2N/mm2}$	N/mm$^{2N/mm2}$	N/mm$^{2N/mm2}$	kg/m$^{3kg/m3}$	N/mm$^{2N/mm2}$	N/mm$^{2N/mm2}$	N/mm$^{2N/mm2}$
Bambusa auriculata	594	65,1	15010	36,7	670	89,1	21410	54,3
B. balcooa	783	65,4	7310	46,7				60,6
B. bambos	559	58,3	5950	35,3	663	80,1	8960	53,4
B. burmanica	570	59,7	11010	39,9	672	105	17810	65,2
B. glancescens	691	82,8	14770	53,9				
B. nutans	603	52,9	6620	45,6	673	52,4	10720	47,9
B. pallida	731	55,2	12900	54				
B. tulda	658	51,1	7980	40,7	722	66,7	10070	68
B. ventricisa	626	34,1	3380	36,1				
B. vulgaris	626	41,5	2870	38,6				
Cephalostachum pergracile	601	52,6	11160	36,7	640	71,3	19220	49,4
Dendrocalamus longispathus	711	33,1	5510	42,1	684	47,8	6060	61,1
Dendrocalamus strictus	631	73,4	11960	35,9	728	119,1	15000	69,1
Melocanna baccifera	817	53,2	11390	53,8	751	57,6	12930	69,9
Oxytenanthera abyssinicia	688	83,6	14960	46,6				

Table 1. Physical and mechanical properties of Indian bamboos (in round form) [14].

Shear stress parallel to grain is 6–12 N/mm^2 which is approximately 10 times lower than compressive strength and up to 20 times lower than the tensile strength of the same bamboo species. However, the shear strength of bamboo is often twice the value of popular timber species. (https://www.bambooimport.com/en/blog/what-are-the-mechanical-properties-of-bamboo). The bending strength of most bamboo species varies between 50 and 150 N/mm^2 (**Table 2**) and is, on average, twice the magnitude of most conventional structural timber materials. Interspecies variations can be caused by different test methods, sample quality and moisture content of the tested bamboo. (https://www.bambooimport.com/en/blog/what-are-the-mechanical-properties-of-bamboo).

When bamboo is used in green condition, the mechanical values that should be used for design are shown in **Table 3** (adapted from [14]).

Laboratory testing of material properties [15, 16] showed that the compressive and shear strength parallel to the grain were most significantly affected by moisture content, followed by longitudinal tensile modulus and then bending modulus. Age had little effect on the sensitivity of the tensile modulus and bending modulus to moisture content change, while young

Category	Modulus of rupture (R) N/mm^2	Modulus of elasticity (E) N/mm^2	Max. compressive strength ($f_{c,max}$) N/mm^2
A	>70	>9000	>35
B	50-70	6000-9000	30-35
C	30-50	3000-6000	25-30

Table 2. Material properties of different categories of structural bamboo (adapted from BIS [14]).

Category	Species	Extreme fibre stress in bending N/mm²	Modulus of elasticity N/mm²	Allowable compressive stress N/mm²
A	Bambusa glancescens	20,7	3280	15,4
	Dendrocalamus strictus	18,4	2660	10,3
	Oxytenanthera abyssinicia	20,9	3310	13,3
B	Bambusa baicooa	16,4	1620	13,3
	B. pallida	13,8	2870	15,4
	B. nutans	13,2	1470	13
	B. tuida	12,8	1770	11,6
	B. auriculata	16,3	3340	10,5
	B. burmanica	14,9	2450	11,4
	Cephalostachum pergracile	13,2	2480	10,5
	Molocanna baccifera	13,3	2530	15,4
	Thyrsostachus oliveri	15,5	2160	13,4
C	Bambusa arundinacea	14,6	1,32	10,1
	B. ventricosa	8,5	0,75	10,3
	B. vulgaris	10,4	0,64	11
	Dendrocalamus longispathus	8,3	1,22	12

Table 3. Safe permissible stresses of different categories of green bamboo for structural design (adapted from BIS [14]).

bamboo was more sensitive to moisture content change for shear strength and less sensitive for compression strength [17].

The experimental results show that the tensile strength of bamboo roots decreases with the increase in diameter through a power function. The tensile strength of the tested bamboo roots ranges between 18 and 30 N/mm² at ultimate strains of 14–18% [18].

4.4. Bamboo natural durability

Bamboo, like most lignocellulosic materials, has very low resistance to biological degrading agents. The culms are liable to attack by insects and termites, and above the fibre saturation point, they can be deteriorated by strain and rotting fungi [19].

Bamboo is more susceptible to decay than timber, due to a lack of natural toxins and its typically thin walls. This means that a small amount of decay can mean a significant percentage change in technical capacity [20].

The high sugar and starch content of the bamboo culm explains its low natural durability. There is not much data available on the natural durability of different bamboo species. The natural durability of raw bamboo is low and varies between 1 and 36 months depending on the species, age of culms and climatic conditions [21]. Bamboo is generally destroyed in about 1–2 years when used in the open and in contact with the ground. According to durability classification [22], bamboo falls in class III (non-durable category) with little variation among species. Data about natural durability of some bamboo species obtained from tests of untreated bamboo poles can be found in [21].

The most popular traditional treatment for enhancing bamboo natural durability is soaking the bamboo in running water for a period of time [19]. This treatment has a significant effect in enhancing durability against decay fungi because it washes off the starch content of the culms [19]. This treatment has little effect on termite attack because these organisms depend merely on cellulose rather than on starch as food source [23]. Traditional techniques for controlling starch content in felled bamboo include [21]:

- Felling of bamboo during low-sugar content season
- Felling of bamboo at maturity when sugar content is low
- Postharvesting transpiration of bamboo culms
- Water soaking of bamboo

When bamboo is used as an inert construction material, seasoning (drying) processes are important in order to carefully bring down its moisture content to levels closer to the in service equilibrium moisture content. Seasoning improves bamboo's resistance to biological attack and limits the amount of drying shrinkage in service [20].

The low natural durability of bamboo material can be perfectly accommodated in the bioengineering approach since, in this type of work, just an initial rigidity is pursued and, hence, only temporary structures are usually included in the work design. Once the introduced living plants become established, the vegetation gradually takes over the structural functions of the wooden supports [24, 25]. As time progresses, the inert bamboo culms will deteriorate, and the live bamboo pegs (or poles) will grow and assume the strengthening effect of the initial structure.

In the bioengineering design approach, the load transfer between the initial structural elements and the evolving structural vegetation elements can be calculated using an eco-engineering design scheme for durability [26]. The rapid growth pattern of bamboo species make them very suitable for this kind of work approaches and strategies.

5. Bamboo effects at the slope stability level

The reinforcement effect ensured by bamboo roots (**Figure 3**) can be expressed in engineering terms as an 'additional cohesion' added to the strength of the non-rooted soil Eq. (1) [27]. Therefore, the total cohesion of a rooted soil will be the sum of the unrooted soil cohesion plus the cohesion increase due to the presence of roots in the soil [27].

The 'additional cohesion' (ΔS; Eq. (1)) can be calculated for a known root tensile strength and root area ratio (RAR; the ratio of the surface area of roots crossing the shear plane and shear plane area [28, 27], Eq. (1)) assuming that all the roots cross the shear plane perpendicularly and break during the shearing process. The rooted soil strength value is then used in traditional slope stability analysis methods (e.g. limit equilibrium methods) to determine the overall slope stability:

$$\Delta S = 1.2\, t_R \ \mathrm{RAR} \tag{1}$$

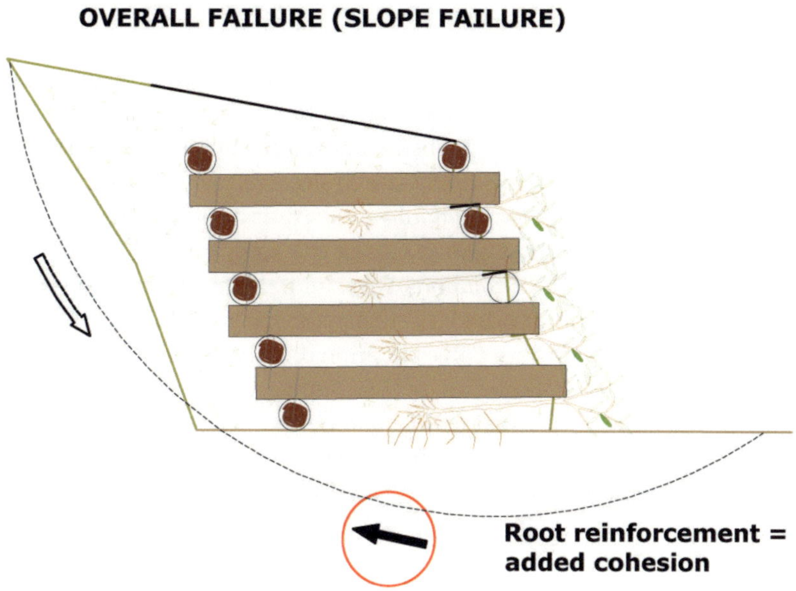

Figure 3. Overall stability check for a bamboo crib wall.

where ΔS is added cohesion or increase in shear strength due to the presence of roots in the soil [KN/m^2], t_R is average tensile strength of roots per unit area of soil [KN/m^2], and RAR is the ratio of area of roots crossing the shear plane and the shear plane area.

The bamboo capacity to improve overall slope stability is limited by its shallow root system. Important effects can be found within the first 0.3–0.5 m depth. Examples of assessment of soil strength increase by bamboo living culms roots can be found in [29].

At the end of the construction stage, when the bamboo culms roots are not yet developed, the slope must be kept stable exclusively by the action of the inert elements and structures used in the bioengineering work. As time progresses, bamboo roots will increase the overall slope stability, and this process can be assessed by using Eq. (1) and traditional slope stability analysis methods.

6. Design standards for bamboo structure calculations

The current bamboo construction standards published in 2004 by the International Organization for Standardization (ISO) were the first step in the attempt to standardize the use of bamboo in construction [15, 16, 30]. These standards are essentially based on the existing traditional knowledge with an adaptation of the existing ISO timber and timber testing standards for bamboo [31]. These standards cover the basis for design and testing of bamboo and bamboo products and can be used as a basis for further standardization of bamboo as a structural material used in soil bioengineering. ISO 22156: Bamboo, structural design [30], provides basic design guidance for full culm bamboo construction. This standard is supported by ISO 22157-1 Bamboo, determination of physical and mechanical properties, part 1: requirements [15], which specify the test methods necessary for design, and ISO 22157-2 [16], which is essentially a laboratory manual for determining the structural properties of bamboo.

6.1. ISO 22156: bamboo: structural design

This international standard is based on limit state design and on the structure's performance; it is only concerned with the requirements for mechanical resistance, serviceability and durability of structures [30]. Execution (work on-site and fabrication of components off site and their erection on-site) is covered to the extent that is necessary to indicate the quality of construction materials and products which should be used and the standard of workmanship on-site needed to comply with the assumptions of the design rules. Bamboo construction design concepts shall be based on calculations, relevant permissible stresses verifying that no relevant limit state or no stress is exceeded. Exclusions are made for design based on previous generations' experience or design based on evaluation reports on structures that survived natural disasters (e.g. hurricanes, earthquakes, etc.) undamaged. In this standard, the limit state design is based on the characteristic value of a material property (5 percentile property, estimated from test results [15], with 75% confidence that it represents the sampled population). The standard advises that special attention is given to differences between materials originating from different sites to account for natural variability. The standard assumes that bamboo will behave as a linear elastic material with service classes dependent on the local environmental conditions (e.g. temperature, humidity) during the structure lifetime. This means that bamboo material is expected to behave elastically until failure, while the plastic behaviour is considered insignificant. In terms of schematization; the bamboo culms are supposed to be analysed as not perfectly straight, tapered, hollow-tube structures with variable thickness.

Similarly as for timber structures, the design shall be verified if no possible limit state is exceeded when partial safety factors and loads/actions relevant to the location of the structure have been applied. Alternative to this approach, allowable stress approaches can be adopted with suitable modification for differences between laboratory and in situ results (0.5), duration of the load (1.0–1.5) and a default value of the factor of safety of 2.25. In these analyses, the conventional structural analysis methods can be used with the bamboo initial curvature, diameter and wall thickness as inputs. The joints/supports of bamboo structures should be located near the nodes (which, in reality, are not spaced at constant intervals) and should be considered to act as a hinge, unless substantive data exist to justify a spring or a fixed joint.

Reflecting the use of bamboo structural members in bioengineering, this standard prescribes the design of beams (predominantly loaded in bending), columns (predominantly axially loaded), joints and assemblies (trusses). All beam elements should be symmetrically loaded, and the loads should be applied preferably close to or at the nodes. All axially loaded bamboo elements should preferably be constructed using the best available straight bamboo culms to avoid buckling. For both types of elements, standard structural calculation methods apply, taking into account the effects of any combination of stresses that may occur during the structure/element lifetime. The joints between the elements should be rigid and provide structural continuity between them, including force transmission and deflection limitation.

In the section on sound construction practices, this standard advises that the designer is in charge of ensuring that 'sound construction practices are taken into account', covering mostly the moisture content change of structural bamboo. It also suggests that special care should be taken to ensure the workmanship (on-site and for products coming from the factory) is according to the assumptions listed in the appendix of the standard. Special provision is made for the use of bamboo elements as soil reinforcement, providing that there is an appropriate evidence (test results) that the bamboo will function as reinforcement during the structure's expected service lifetime, with special attention to the lifetime of the bamboo in the organic environment. Types of tests and the frequency of testing are not specified in this standard.

The environmental conditions (temperature, humidity, moisture content, soil/water characteristics and composition, surrounding flora/fauna, etc.) must be taken into account at the design stage to assess their effect on the durability of the structure but also to enable the design of material protection techniques. The durability of each structure is expected to vary based on the particular materials and the environmental conditions. The standard does not prescribe methods for assessing the durability but suggests considering a range of factors (e.g. environment, service life, use, performance criteria, workmanship, maintenance, etc.) when making such assessments.

In terms of quality assurance, the design with bamboo structural elements should be carried out by suitably qualified and experienced personnel. Similarly, qualified and experienced personnel should carry out the supervision and quality control during construction. The structures should be used as per the design briefly and adequately maintained. This standard specifies quality control for mainly factory-produced bamboo and products using a quality assurance manual (QAM). For the purposes of eco-engineering works, the standard can be interpreted including material specifications (including incoming material, inspection, and acceptance requirements), quality assurance inspection testing and acceptance procedures, sampling and inspection frequencies and procedures to be followed upon failure to meet specifications or upon out-of-control conditions. The QAM should be supplemented by relevant records, including inspection and test records, test data, corrective actions, etc.

6.2. ISO 22157-1 bamboo: determination of physical and mechanical properties: part 1: requirements

This standard specifies the test methods for evaluating the following characteristic physical and strength properties of bamboo: moisture content, mass per volume (density), shrinkage, compression, bending, shear and tension. This data is needed for establishing characteristic strength functions and determination of allowable stresses, as well as for establishing the relationship between mechanical properties and factors for quality control functions.

The standard prescribes the acceptable precision of testing measurements, the sampling and storage of test samples and reporting requirements. The measurements specified in the standard include determination of moisture content, mass (by volume), shrinkage, strength in compression (including nominal modulus of elasticity), strength in bending (including load-deflection curves and modulus of elasticity), strength in shear and strength in tension. The principles, apparatus, preparation of test specimens, procedure, calculation and expression of results and test report requirements are specified for each test.

6.3. ISO 22157-2 bamboo: determination of physical and mechanical properties: part 2: laboratory manual

The purpose of this standard, originally written as a technical report is to disseminate best practice test methods in order to make these globally available but also to outline the 'how to' for the tests specified in ISO 22157-1 [15, 16]. On a number of issues, this standard refers to existing national and supranational standards which should be followed, especially for bamboo application in temperate climates (e.g. W. Europe, Canada). This standard contains a number of examples, templates and backgrounds to the tests specified in [15].

The above ISO standards provide the basis for design with bamboo culms. However, they do not include content relating to living bamboo (design and testing) and the use of a combination of living and inert bamboo in a structure. The existing standards need to be updated and expanded to reflect the growing research on test methods and material characterization of both living and inert bamboo, especially covering the durability and evolution of load transfer mechanisms with time. This update could be partially covered by an attempt to use the existing timber-based test methods for characterization and design [29] which could be beneficial in engaging engineers and architects [31] in the use of bamboo for bioengineering purposes.

7. Stability checks for bamboo retaining structures

The analysed structures can be classified as soft engineering structures [24] with a certain durability and change in stress transfer mechanism over the lifetime of the structure. The durability of the structure will depend on the used bamboo species and the biological activity of local degrading mechanisms but also on air temperature, humidity and soil moisture variability. In the bioengineering design approach, the load transfer between the initial structural elements and the developing structural vegetation elements can be calculated using a bioengineering design scheme for durability [26]:

- Determination of the mechanical properties of the wooden elements
- Determination of the stress diagrams of the different structural elements
- Determination of the decay rate of the wooden element and their design service life
- Determination of the plant root system growth and the roots' mechanical properties
- Stability assessment of the structure at different periods of its design lifetime reflecting the progression of decay and development of the live elements in the structure [26]

7.1. External stability checks

As with any stabilization structure, soil bioengineering solutions must be checked from a structural point of view to ensure that both external (sliding, overturning, bearing capacity and slope failure; **Figure 1**) and internal stability conditions are satisfactory. These checks must include both decay and living plant effects, in order to reflect the changes during the lifetime of the bioengineering solution. The external stability checks are usually performed in line with existing geotechnical engineering design standards and the stability is expressed in terms of a factor of safety (FoS; e.g. [32]). In this book chapter, both the FoS expressions for bare and vegetated soil [24] and the use of lumped global FoS for the sliding and overturning checks are proposed. The resistance to sliding (FoS_s) will be affected by the evolution of the RAR value with time across the sliding plane [33], while the resistance to overturning (FoS_o) will be affected by the pull-out force evolution with time due to root growth (**Figure 4**). As shown before, the overall slope stability of a bioengineered slope can be assessed using existing slope stability analysis methods [34] taking into account both long-term (drained) and short-term (undrained) conditions.

7.2. Internal stability check

Internal stability analysis consists of checking the mechanical capacity of the bamboo culms which are fulfilling structural functions within the ground bioengineering work.

The characteristic strength values should be obtained according to ISO 22157 [15, 16]. Suggested characteristic strength values for any bamboo species can be found in [35]. These values should be adapted to the bamboo moisture content by using the moisture content correction factor included in [35] which is based on NSR [36] and EN 384 [37]. Bamboo live pole (pegs) strength characteristic values have been shown in preceding epigraphs (see **Table 3**). The values for the material factor of safety are specified in ISO 22156 [30]. Recommended values for this factor can also be found in [35].

In soil bioengineering works, bamboo culms work under Service Class 3 conditions (relative humidity >85%) and in-ground conditions. This situation can be reflected by means of the service class and load duration factor (K_{mod}) which can be determined from the existing standards [30, 35].

Other factors making allowance for other conditions (e.g. earthquakes, connection between elements of different rigidity, etc.) can also be found in [35].

Bamboo structures' internal stability checks follow Eurocode 5: Design of timber structures. This design scheme is detailed in ISO 22156 [30] and in [38]. Because of the specific bamboo culm shape, the cross-sectional area to be used in the design calculation is the following:

$$A = \frac{\Pi}{4}(De^2 - (De - 2t)^2) - \sum area\ of\ any\ holes \tag{2}$$

where A is the net area of section; De is the bamboo culm outer diameter; t is the bamboo culm average thickness.

Accordingly, the elastic section modulus ($S_{elastic}$) for the bamboo case will be.

$$S_{elastic} = \frac{\Pi(De^4 - (De - 2t)^4)}{32De} \tag{3}$$

By comparing the existing stress values with the element bending capacity, shear, axial tension or axial compression, the internal stability check can be fulfilled.

Additionally, Eqs. (2) and (3) can also be useful for determining the minimum bamboo culm dimeter fulfilling the internal stability condition [29].

The forces exerted on the bamboo culms can be determined by using traditional structural calculation theory (**Figure 5**). Typically, the moment, shear and axial stress diagrams should be generated and their maximum values used for internal stability checks [29].

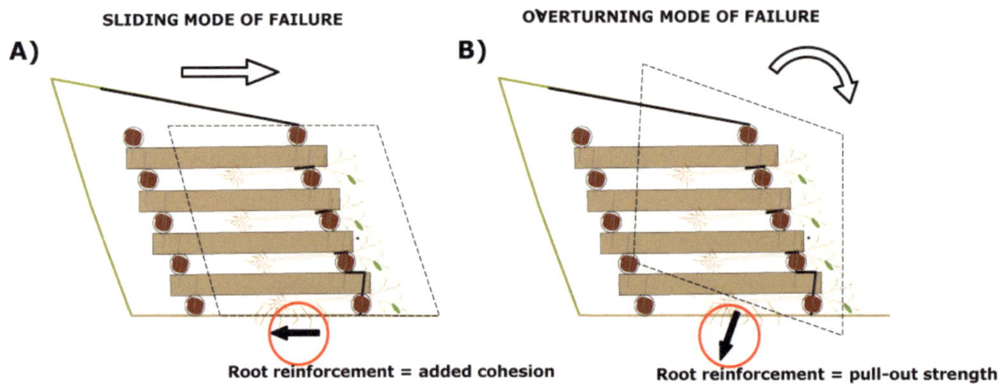

Figure 4. (A) Sliding check and (B) overturning check. The bamboo root effect (if applicable) is highlighted within the circle (adapted from [26]).

Given the low natural durability value of bamboo species, an adapted design scheme making allowance for the wooden element deterioration process can be adopted [26]. In order to give answer to cross-sectional losses because of decay processes at the internal stability design level, different strategies can be followed:

- Increase of the initial diameter of the bamboo culms used in the work
- Reduction of the elements bending span length
- Reduction of the height of the retaining structures

In the last two cases, lower forces will be exerted on the bamboo culms, and, therefore, lower diameters will be able to withstand them and ensure the internal stability of the bamboo structure. A complete example of the preceding design approach can be found in [26, 29].

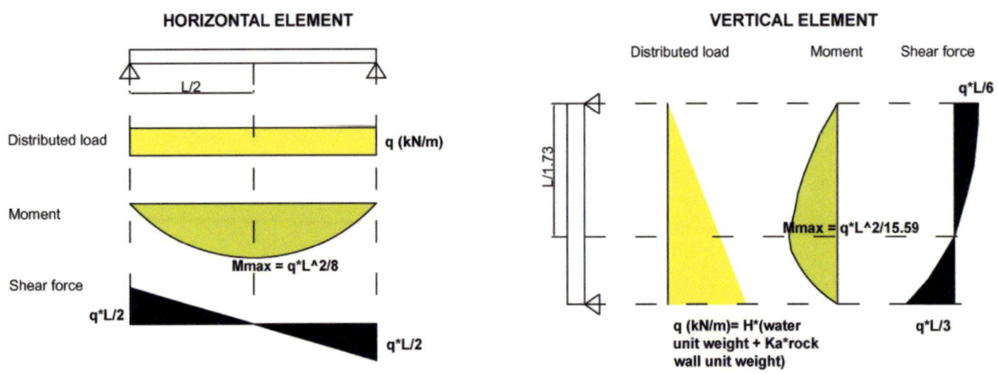

Figure 5. Example of stress diagrams for the vertical and horizontal elements in a bamboo palisade [29], where q is the distributed load (kN/m^2), L is the length of the element (m), H is the height of the vertical element and Ka is the coefficient of active earth pressure (dimensionless).

8. Plant material quality control, maintenance and monitoring of bamboo structures in soil bioengineering

Quality assurance manual (QAM) covering the bamboo application and approved by the overseeing organization should form an integral part of soil bioengineering project documentation [30]. For the purposes of soil bioengineering works, the QAM should record the control of plant material (including incoming material, inspection and acceptance requirements), personnel, design, construction, quality assurance testing and acceptance procedures, sampling and inspection frequencies and procedures to be followed upon failure to meet specifications or upon out-of-control conditions. The QAM should be supplemented by relevant records, including inspection and test records, test data, corrective actions, etc.

8.1. Plant material quality control

There are a large number of species of bamboo native to different world regions, mainly in the warm and moist tropical and warm temperate climates. Current standards recognize that methods of identification of bamboo through anatomical characteristics have not been perfected. To mitigate against the risk of selecting inadequate/untested species, experienced sorters should be employed in identification through morphological characteristics on full standing culm and the results recorded in the QAM.

Quality control must ensure that only matured bamboo of at least 4 years of age shall be used in construction, preferably after at least 6 weeks after felling. Experienced quality assessors should ensure that solid bamboo culms or culms with thicker walls and closely spaced nodes are selected for structural use. Conversely, the quality assurance control must ensure that any broken, damaged or collapsed bamboo shall be rejected while dead/immature/infected bamboos shall be avoided.

If living bamboo is to be used in the structure, the origin/provenance of the bamboo seeds/seedlings/plantings should be recorded in the QAM together with the species name(s), application/planting rate/density, fertilizer, mulching, soil preparation and maintenance requirements. This must be supported by certificates relating to type, origin, quality and validity of seeds/plants and quality of fertilizers.

Knowing that the natural durability of bamboo is relatively low (12–24 months when used in the open and in contact with the soil but depending on the species and environmental conditions), and its strength decreases rapidly with the onset of fungal decay, suitable treatment (traditional treatments if possible) for preserving bamboo must be applied considering the environmental impact and health aspects of labour and all users of the structure. The samples for testing the effect of preservatives must be cut from treated bamboo for chemical analysis (e.g. a weight of approx. 100 g per 100 kg bamboo treated).

Air-dried bamboo should be used whenever possible in order to ensure it deteriorates more slowly. QC should ensure that, if the bamboo delivered to site is wet, there is an opportunity to dry again before it is applied in the soil bioengineering structure.

8.2. Personnel

Bamboo structure should be designed and constructed by personnel having appropriate skills and experience. Similarly, qualified and experienced personnel should carry out the supervision and quality control during construction. The expertise and skills of the personnel involved must be recorded at an early stage as part of the quality management submission of the contractor and then updated periodically throughout the duration of the project.

8.3. Design

Design tasks (e.g. load analysis, calculations, specifications, drawings, detailing) and/or modelling should be carried out as per the existing current standards (see Section 6 above) and the design brief. Here it is acknowledged that traditional experience rather than precise calculations may generally govern the detailing but the experience has to be based on evidence in form of reports on the structural damage to similar structures after they have sustained the severity of earthquakes, hurricanes, etc. as criteria for recommendations by the evaluation by competent engineer/builder with adequate experience in the field. Independent design check should be carried out in accordance with the quality management demands and methods of the client and the contractor.

8.4. Laboratory and in situ testing

Laboratory and in situ testing of material properties, structural stability and general performance of the structure should be carried out in accordance with the existing current relevant standards (see Section 6) and design brief [15, 16]. The quality control should ensure that due care is taken to maintain and calibrate the testing equipment on a regular basis and minimize the scatter in the test data by the use of appropriate numerical techniques, a representative number of tests is carried out for determination of each property and that the personnel performing the tests and the analysis of results are suitably qualified and skilled. The records covering the above should be kept as part of the QAM and periodically updated throughout the duration of the project.

8.5. During the construction stage

The quality control should ensure that the workmanship of the labour force is according to the design and testing assumptions. Regular periodic inspections should be carried out by a designated quality assessor or manager who will record any defects stemming from the workmanship and allow the project manager or site supervisor suitable time to correct them before proceeding. The inspection records should be kept as part of the QAM, with a special emphasis of the following defects when using inert bamboo culms:

- Damage caused by bamboo Borer/ghoon beetle (*Dinoderus* spp. *Bostrichidae*), which attacks felled culms.
- Crookedness of the bamboo culms in terms of a localized deviation from the straightness in a piece of bamboo.

http://dx.doi.org/10.5772/intechopen.75626

- Discoloration of the bamboo demonstrated as a change from the normal colour of the bamboo which does not impair the strength of bamboo or bamboo composite products.
- Collapse of the culm occurring on account of excessive shrinkage, particularly in thick-walled immature bamboo, and causing a reduction in structural strength. This defect is demonstrated as a development of a wide split resulting in a depression on the outer surface of the culm.
- Splitting at the end of a bamboo as an effect of drying which occurs both from outer and interior wall surfaces of bamboo as well as the end at the open ends.
- Surface cracking demonstrated as fine surface cracks not detrimental to strength which can reduce the structural strength if it occurs at the nodes.
- Wrinkles and deformation in cross section, during drying, which occurs in immature round bamboos of most species; in thick-walled pieces, besides this deformation the outer surface becomes uneven and wrinkled. Very often the interior wall can develop a crack below these wrinkles, running parallel to the axis and decreasing the structural strength of the culm
- If the engineer has identified a **fire risk**, the QA should check if the necessary treatment has been applied to the bamboo prior to construction/commission.

The client should ensure that the structure will be **used and operated** in accordance with the design limit briefs. Periodic inspections and monitoring of the use of the structure and the loads experienced should be carried out and the records kept in the QAM during the lifetime of the structure. Emergency action plan should also be prepared by the designer/engineer in order to cover the mitigation measures to be put in place during or after the application of accidental loads or in the event of structural failure.

8.6. Maintenance

The existing standards [30] specify that adequate maintenance is required for the structure without providing definition of adequacy and the quality control procedures associated with it. Knowing that one of the main characteristics of soil bioengineering works is that their full efficiency is only reached once the living plants have rooted and active growth has commenced [39], it is critically important that the maintenance is planned and controlled in order to accelerate the establishment process and shorten the time between construction and reaching full operational capacity. In this respect, the quality control should cover maintenance during the plant establishment phase (recording the percentage of ground cover, application of fertilizers and mulches, survival/failure rate of plantings/seedlings, inspection and acceptance/rejection of all live and inert plant materials), aftercare during the plant establishment phase (replacement of inadequate/dead plant material, fertilization, weeding, cultivation, mulching, irrigation, protection/preservation, staking/tying) and aftercare during development stage (periodic inspections during 2–5 years after construction and, if needed, replacement of any inadequate materials, fertilization, irrigation, ground preparation, mulching, mowing, pruning, staking/tying, pest disease control [14, 40], coppicing and pollarding).

8.7. Monitoring of bamboo structures and applications

As a result of working with materials possessing relatively large natural variability, the soil bioengineering design with bamboo has to account for some uncertainty. To mitigate against the uncertainties and risks, monitoring will be vital throughout the lifetime of the bamboo-based structure. Current standards do not specify or regulate which monitoring should be conducted as a step-by-step engineering process starting with a definition of objectives and end with application of mitigation measures if warranted by the monitoring data [41]. A monitoring programme should be included in the design and periodically updated as part of the QAM in the project file throughout the duration of the project. The programme should detail the specification of instrumentation and methods for monitoring and should be cross-referenced to the risk assessment register of the project.

The monitoring of a soil bioengineering structure can be carried out in accordance with the existing current guidelines (e.g. [42]) with an addition of monitoring of the bamboo-related parameters critical to the stability and resilience of the structure. Depending on the form of bamboo used (live or inert) in the structure, the monitoring of the following parameters should be adopted as a minimum:

- Survival rate and/or percentage ground cover
- Height above ground or length of culms
- Displacements (horizontal, vertical, tilt) of the bamboo elements and the whole structure
- Root spread (horizontal and vertical)
- Root density
- Moisture content (air and soil)
- Temperature
- Culm diameter range
- Tensile strength of the roots
- Suction stress due to the presence of roots
- Groundwater levels
- Changes in soil organic content
- Precipitation
- Runoff
- Soil loss to erosion
- Diameter deterioration and rates of deterioration
- Bamboo strength [15, 16]

A sustainability assessment framework can be used to benchmark the sustainability performance of the bamboo-based soil bioengineering applications for the purpose of monitoring their performance. In this context, an attempt to cover the socioeconomic, environmental and engineering performance of the application can be performed, and the framework can be used as part of the existing QA/QM procedures with a number of KPIs already being measured as part of the other QA processes [43]. The advantage of such a framework would be its use to assess the performance of the application after construction. The graphical output of the assessment makes the framework easier to use also throughout each project stage as a planning and decision-making tool.

9. Case studies: analysing accumulated experiences

The following shows a case study conducted under a research project from the University of Natural Resources and Life Sciences Vienna (BOKU) and the Tribhuvan University Kathmandu, Nepal [44].

The research work was focused on field investigations to develop technical standards of soil bioengineering systems. Among other research activities, one site in Kusunti in Kathmandu was selected for the implementation of a bamboo crib wall. A vegetated bamboo crib wall was compared with a conventional slope stabilization method (gabion) by means of different parameters.

At Kusunti site, that half portion of the site was treated with the gabion retaining wall, and the other half with the bamboo crib wall to compare two retaining wall systems from a technical as well as an economic point of view. One layer of gabion retaining wall was constructed for the whole stretch as a base. The total designed height of the wall is 3 m.

The actual construction work at this site was started on November 11, 2006 and ended on November 23, 2006. A supervisor was appointed to control the quality of the work. Students from the Pulchowk Campus and one student from Switzerland (University of Applied Sciences Wädenswil (HsW)) were also directly involved in this work. The before and after construction photographs and the work evolution are shown in **Figures 6** and **7**, respectively.

From the various project activities and critical study and monitoring of project sites, the following general conclusions on the use of bamboo crib walls can be made:

- Vegetative crib walls are comparatively cheaper than gabion or stone masonry wall (construction costs only ¼ of gabion and 1/5 of masonry wall) but provide the same technical stability.
- Although there is more vertical settlement in the case of vegetative crib wall, compared to gabion/masonry walls, vegetative crib walls have better attachment with the slope.
- Vegetative crib walls keep minimum soil moisture, avoid cracks in soil and provide better interception during rainfall.

Figure 6. Before and after construction photographs, Kusunti, Lalitpur (November 2007) [26].

Figure 7. The vegetated bamboo crib wall just after construction (November 2006) and after 1 year of construction (Kusunti, Lalitpur, November 2007) [26].

- More stability through increased cohesion and soil reinforcement is provided in the case of vegetative crib walls.
- The vegetative crib wall made of wood or bamboo is more suitable (also sustainable and environmentally friendly alternative) for solving slope stability problems or for road embankment protection in Nepal.

10. Conclusions

The existing accumulated experiences of using bamboo in soil and water bioengineering works, together with the existing standards and design guidelines, make specific bamboo species an essential and cost-effective material for erosion control and slope stabilization works where these species are native.

The integration of the bioengineering particularities at the design stage demands the integration of the plant evolution and the deterioration processes into the bioengineering design scheme. The existing design routines can be adapted for making allowance of the preceding features. The analysis of other works, the accumulated experiences of monitoring and field works and tests (e.g. bamboo root depth, root tensile strength, bamboo culm deterioration processes, bamboo culm mechanical testing, etc.) will support the specialization process for this type of interventions in the future.

Author details

Guillermo Tardio[1]*, Slobodan B. Mickovski[2], Hans Peter Rauch[3], Joao Paulo Fernandes[4] and Madhu Sudan Acharya[3]

*Address all correspondence to: gtarcer@gmail.com

1 Technical Univesity of Madrid, Spanish Association of Landscape Engineering, Madrid, Spain

2 Glasgow Caledonian University, Glasgow, Scotland, UK

3 Department of Civil Engineering and Natural Hazards, University of Natural Resources and Life Sciences, Vienna, Austria

4 University of Evora, Evora, Portugal

References

[1] Studer R, Zeh H. Soil Bioengineering: Construction Type Manual. Zurich: vdf Hochschulverlag AG; 2014

[2] Florineth F. Pflanzenstatt Beton. Handbuch zur Ingenieurbiologie und Vegetationstechnik. Berlin: Patzer Verlag; 2012

[3] Hacker E, editor. European Guidelines for Soil and Water Bioengineering. Aachen: European Federation for Soil Bioengineering; 2015

[4] Sthapit KM, Tennyson LC. Bio-engineering Erosion Control in Nepal. Rome, Italy: Food and Agriculture Organization of the United Nations; 2014. Available from: http://www.fao.org/docrep/u1510e/u1510e04.htm

[5] Morgan RPC, Rickson RJ. Slope Stabilization and Erosion Control: A Bioengineering Approach. London, UK: E&FN Spon; 1995. pp. 1-47 and 221-248

[6] Lammeranner W, Rauch HP, Laaha G. Implementation and monitoring of soil bioengineering measures at a landslide in the Middle Mountains of Nepal. Plant and Soil. 2005;**278**:159-170

[7] Rauch HP, Florineth F, Lammeranner W, Wibmer S. Soil Bioengineering Slope Protection Investigations in Nepal. Final Report for the Austrian Commission of Development Research. 2002

[8] Scurlock JMO, Dayton DC, Hames B. Bamboo: An overlooked biomass resource? Biomass and Bioenergy. 2000;**19**(4):229-244

[9] Banik RL. Bamboo silviculture. In Bamboo. Springer International Publishing; 2015. pp. 113-174. DOI: 10.1007/978-3-319-14133-6

[10] Stokes A, Sotir R, Chen W, Ghestem M. Soil bio-and eco-engineering in China: Past experience and future priorities. Journal of Ecological Engineering. 2010;**36**(3):247-257

[11] Bystriakova N, Kapos V. Bamboo diversity: The need for a red list review. Biodiversity. 2006;**6**(4):12-16

[12] Rogerson A, Edi Suhaimi B, Jegatheswaran R, Awang KB. Bamboo properties and suitability as a replacement for wood. Pertanika Journal of Scholarly Research Reviews. 2016;**2**(1):63-79. eISSN: 2462-2028 © Universiti Putra Malaysia Press

[13] Canavan S, Richardson DM, Visser V, Le Roux JJ, Vorontsova MS, Wilson JR. The global distribution of bamboos: Assessing correlates of introduction and invasion. AoB Plants. 2016;**9**(1):plw078

[14] BIS. CED46(8025)WC – Draft National Building Code of India, Part 6 Structural Design, Section 3 Timber and Bamboo, 3B Bamboo (First Revision). New Delhi, India: BIS; 2015

[15] ISO 22157-1: (E): Bamboo – Determination of Physical and Mechanical Properties – Part 1: Requirements. Geneva, Switzerland: ISO; 2004

[16] ISO 22157-2: (E): Bamboo – Determination of Physical and Mechanical Properties – Part 2: Laboratory Manual. Geneva, Switzerland: ISO; 2004

[17] Jiang Z, Wang H, Tian G, Liu X, Yu Y. Sensitivity of several selected mechanical properties of moso bamboo to moisture content change under fiber saturation point. In: Proceedings of the 55th International Convention of Society of Wood Science and Technology, August 27-31, 2012, Beijing, China. 2012

[18] Hui S, Zhang YW. Study on the tensile mechanical properties of Sympodial bamboo single root. In: Proceedings of the 2012 International Conference on Mechatronics and Computational Mechanics (ICMCM 2012). Stafa-Zuerich: Trans Tech Publications. ISSN 16609336. Ser.no.: 307. pp. 421-426

[19] Zaidon A, Nazri M. Traditional treatment of Malaysian bamboos: Resistance towards white rot fungus and durability in service. Pakistan Journal of Biological Sciences. 2000;**3**(9):1453-1458, 2000

[20] Kaminski S, Lawrence A, Trujillo D, King C. Structural use of bamboo – Part 2: Durability and preservation. In: The Structural Engineer. Journal of the Institution of Structural Engineer. October 2016;**94**(10):38-43 ISSN 1466-5123

[21] Kumar S, Shukla KS, Dobriyal PB. Bamboo Preservation Techniques: A Review. New Delhi: International Network for Bamboo and Rattan (INBAR) and Indian Council of Forestry Research Education (ICFRE); 1994. pp. 59

[22] Anon. Code of Practice for Preservation of Timber IS:401. Manak Bhawan, New Delhi; Bureau of Indian Standards; 1982

[23] Sajap AS, Amit S, Welker J. Evaluation of hexaflumuron for controlling the subterranean termite *Coptotermes curvignathus* (Isoptera: Rhinotermitidae) in Malaysia. Journal of Economic Entomology. 2000;**93**:429-433

[24] Gray DH, Sotir RB. Biotechnical and Eco-Engineering Slope Stabilization. New York: Wiley; 1996. p. 276

[25] Lammeranner W. Slope stabilization with soil-bioengineering methods in the Middle Mountains of Nepal. Suitability of plants for drainage fascines, palisades, single tree planting and grass planting in winter plantation [Master thesis]. University of Natural Resources and Live Sciences, Vienna, Austria. 2002

[26] Tardio G, Mickovski SB. Implementation of eco-engineering design into existing slope stability design practices. Ecological Engineering. 2016;**92**:138-147

[27] Wu TH, Mckinnell WO, Swanston DN. Strength of tree roots and landslides on prince-of-Wales-Island. Alaska. Canadian Geotechnical Journal. 1979;**16**(1):19-33

[28] Waldron LJ. Shear resistance of root-permeated homogenous and stratified soil. Soil Science Society of America Journal. 1977;**41**(5):843-849

[29] Tardio G, Mickovski SB, Stokes A, Devkota S. Bamboo structures as a resilient erosion control measure. In: Proceedings of the Institution of Civil Engineers – Forensic Engineering. 2017;**170**(2):72-83

[30] ISO 22156: 2004 (E): Bamboo – Structural Design. Geneva, Switzerland: ISO; 2004

[31] Gatoo A, Sharma B, Bock M, Mulligan H, Ramage MH. Sustainable structures: Bamboo standards and building codes. Engineering Sustainability. 2014;**167**(ES5):189-196

[32] Tardio G, Mickovski SB. Method for synchronisation of soil and root behaviour for assessment of stability of vegetated slopes. Journal of Ecological Engineering. 2015; **82**:222-230

[33] Preti F, Cantini C. Evoluzione temporale delle condizioni di stabilità per le palificate vive. In: proceedings of Atti XXVIII Convegno di Idraulica e Costruzioni Idrauliche, Potenza, Italy; September 16-19, 2002; Provincia di Teramo, Atti Convegno: Opere in grigio, opere in verde. 2002

[34] Duncan JM, Wright SG. Soil Strength and Slope Stability. Hoboken, New Jersey, EEUU: John Wiley and Sons, Inc; 2005

[35] Kaminski S, Lawrence A, Trujillo D, Feltham I, Felipe López L. Structural use of bamboo. Part 3: Design values. The Structural Engineer. 2016;**94**(12):42-45

[36] Asociación Colombiana de Ingeniería Sísmica (AIS). NSR-10: Reglamento Colombiano de construcción sismo resistente. Titulo G: Estructuras de madera y estructuras de guadua. AIS: Bogota; 2010

[37] CEN. EN 384:1995 Structural Timber – Determination of Characteristic Values of Mechanical Properties and Density. Brussels: CEN; 1995

[38] Kaminski S, Lawrence A, Trujillo D, Feltham I, Felipe López L. Structural use of bamboo – Part 4: Element design equations. In: The Structural Engineer. Journal of the Institution of Structural Engineer. March 2017;**95**(3):24-27

[39] Schiechtl HM, Stern R. Handbuch für naturnahen Erdbau. Eine Anleitung für ingenieurbiologische Bauweisen. Agrarverlag, Vienna: Österr; 1992

[40] BIS. IS9096: 2006 Preservation of Bamboo for Structural Purposes – Code of Practice. New Delhi, India: BIS; 2006

[41] Duncliff J, Marr WA, Standing J. Principles of geotechnical monitoring. In: Burland J, Hapman T, Skinner H, Brown M, editors. ICE Manual of Geotechnical Engineering. London, UK: Institution of Civil Engineers; 2012. pp 1363-1377

[42] BS EN 1997-1. Eurocode 7 – Geotechnical Design Part 1 – General rules. London, UK: BSI; 2007

[43] Mickovski SB, Thomson CS. Developing a Framework for the Sustainability Assessment of Eco-engineering Measures. Journal of Ecological Engineering. 2017;**109**(B):145-160. DOI: 10.1016/j.ecoleng.2017.10.004

[44] Rauch HP, Acharya MS, Khadka P, Shakya NM. Assessment of Soil Bioengineering and Conventional Methods Used in Road Side Slope Stabilisation Work in Nepal (2008) Final Report for the Austrian Commission of Development Research. 2006

Chapter 8

Review of the Resources and Utilization of Bamboo in China

Weiyi Liu, Chaomao Hui, Fang Wang,
Meng Wang and Guanglu Liu

Additional information is available at the end of the chapter

http://dx.doi.org/10.5772/intechopen.76485

Abstract

China has made a breakthrough in the development and scientific cultivation of bamboo. At present, China ranks first in bamboo research worldwide, because of numerous research units and strong technical force. This chapter focuses on the utilization of bamboo resources such as food, roofs and walls of houses, fences, and domestic and agricultural implements such as water containers, food and drink container hats, arrows, quiver, etc. A total of 861 species and infraspecific taxa belonging to 43 genera have been reported and include 707 species, 52 varieties, 98 forma, and 4 hybrids, which are naturally distributed in 21 provinces. The national bamboo forest covers 6.01 million ha, including 4.43 million ha of Moso bamboo and 1.58 million ha of other bamboo species. As the country develops and new economic activities emerge, bamboo production has shifted from harsh processing, such as bamboo basket, to finished machining, such as bamboo flooring. The bamboo industry has attracted new opportunities as a new energy source, particularly renewable energy, and may be considered a lignocellulose substrate for bioethanol production because of its environmental benefits and high annual biomass yield.

Keywords: China, bamboo resources, utilization, Moso bamboo

1. Introduction

Bamboo belongs to the subfamily Bambusoideae and may be grown in any temperate climate zones in Asia, Africa, and Latin America [1]. The total area of bamboo forests is 22.0 × 10 ha, which account for about 1% of the total global forest area [2]. Over the last 3 decades, bamboo has evolved from being a raw material for basic goods into a material base of an increasingly diversified array of products; recently, bamboo has been recognized as a potentially important source of cultural and environmental services [3]. Bamboo species, particularly the large clump bamboo, have enormous potential as an energy source. The bamboo industry plays an important role in economy and society development in China and other countries worldwide [4]. With the expansion of bamboo forests, their importance in people's lives has also increased. Hence, scholars have focused on developing bamboo forest production, increasing the bamboo forest yield, and establishing methods for rational use of bamboo resources in global forestry production.

China has the most abundant bamboo resources worldwide and the richest bamboo culture. Bamboo greatly influences not only the Chinese history and culture but also people's daily life. Although the total forest area in many countries has decreased drastically, the area of bamboo forests has consistently increased at a rate of 3% annually [5].

In recent years, significant advances have been achieved in bamboo cultivation and development, carbon fixation and storage, and ecological and environmental functions in China. This chapter reviews the traditional utilization methods of bamboo resources in China.

2. Basic situation of bamboo resources in China

Bamboo is a rapidly developing, renewable, widely distributed, low-cost, environment-improvement resource. It has the potential to improve poverty alleviation and environment conservation [6]. Bamboos are plants belonging to the subfamily Bambusoideae. According to the data collected, there are 116 genera and 1439 species worldwide [7], of which 62% are native to Asia [8]. China has the most abundant and diverse bamboo species. In China, a total of 861 species and infraspecific taxa in 43 genera have been reported, including 707 species, 52 varieties, 98 forma, and 4 hybrids. Bamboo is naturally distributed in 21 provinces (municipalities or Autonomous Regions) [9]. These taxa include 10 genera and 48 species of endemic bamboo [10]. According to the Eighth National Forest Inventory from 2009 to 2013, statistics shows that the national bamboo forest area covers 6.01 million ha, including 4.43 million ha of Moso bamboo and 1.58 million ha of other bamboo species.

The bamboo forests in China are mainly distributed in 18 provinces. The major provinces with more than 300,000 ha of bamboo forest are Fujian, Jiangxi, Zhejiang, Hunan, Sichuan, Guangdong, Guangxi, Anhui, and the eight provinces (autonomous regions), accounting for 89.10% of the total bamboo forest area. From 1980 to 2013, the bamboo forest area in each province increased annually (**Table 1**).

It is worth noting that the area of bamboo forest resource distribution in Yunnan is not large in China, but Yunnan is considered as one of the original bamboo centers not only in Asia but also worldwide. In Yunnan, 28 genera and at least 250 bamboo species have been found, half of which are of ancient origin [11]. Yunnan has 26 minority nationalities, and people of all nationalities widely use bamboo in clothing, food, shelter, transportation, and in all its aspects in the long-term practice. Bamboo has a profound and unique influence on life and China has a unique special ethnic bamboo culture.

Provinces	1980	1995	2005	2013	Provinces	1980	1995	2005	2013
Fujian	56.71	64.08	91.07	106.75	Guangxi	8.60	13.80	16.32	34.09
Jiangxi	58.20	70.93	93.47	99.89	Guizhou	2.40	3.87	4.43	15.69
Hunan	53.30	68.35	88.80	77.83	Jiangsu	1.80	2.23	2.87	3.37
Zhejiang	45.20	62.45	72.60	83.34	Sichuan	1.60	1.90	2.60	54.90
Anhui	15.30	20.34	24.53	33.72	Taiwan	0.50	0.63	0.80	—
Guangdong	11.27	31.80	17.80	44.62	Yunnan	0.30	—	0.48	11.04
Hubei	8.60	8.60	13.80	14.40	Henan	0.29	0.47	1.67	2.74

Table 1. Statistics of increase in bamboo resources in major provinces and autonomous regions in China (unit: 10,000 ha).

3. Utilization types of bamboo resources in China

Bamboo plays an important role in the development of Chinese historical culture. Through the ages, bamboo has been gaining an increasing attention from Chinese people. As one of the fastest growing plants, bamboo has become one of the most important forest resources and has a high commercial value and ecological function. Bamboo is indispensable in the manufacturing industry and daily life of Chinese people.

China is a developing country that has a large number of population dependent on agriculture. Many of their production equipment and supplies are made of bamboo, especially in the southern rural areas.

Since ancient times, bamboo has been widely used in the daily life of people in rural China. Therefore, most of the bamboos produced in China are used in rural and agricultural production. Bamboo resources in China were traditionally utilized by local farmers as minor forest products with weak linkage in the market. In recent years, with the fast pace of modernization, bamboo plants have replaced wood because of their advantages, such as fast growth, high yield, short rotation, and easy management. The bamboo processing industry has gathered momentum and entered a new stage. By the end of the twentieth century, dozens of countries established bamboo-based paper milling, with annual output of more than 1 million tons of bamboo paper and bamboo pulp production capacity of 4 million tons. Fine processing of bamboo has also gained increasing attention to produce a new type of artificial bamboo board with features similar to those of wood. Research and production of bamboo products have reached the practical level and production scale. The wood-based panel industry has been in the forefront in terms of variety and scale worldwide. Furthermore, the quality and process of mechanical production of bamboo have continuously improved.

Table 2 shows the methods for utilizing bamboos. The main application of bamboo in China is divided into two parts: economic use and ecological utilization. Economic utilization can be roughly divided into timber bamboo, shoots bamboo, skin bamboo, and art and crafts bamboo. Ecological value can be divided into water conservation forest and ecological forest tourism.

Value	Utilization	Specific purpose	Bamboo species
Economic	Timber	Bamboo flooring Pulp and paper Construction material Bamboo chopsticks	*Phyllostachys edulis* *Dendrocalamus giganteus*
	Shoot	Fresh shoot Drying shoot Canned shoots Flavored Shoot	*Ph. edulis* *D. brandisii* *D. latiflorus* *Ph. praecox*
	Skin	Tables and chairs Basket Wall of house	*Neosinocalamus affinis* *Bambusa textilis* *B. chungii*
	Arts and crafts	Musical instruments Bonsai Root carving	*Qiongzhuea tumidissinoda* *Chimonobambusa quadrangularis* *Pseudosasa amabilis* *Ph. nigra*
Ecological	Water conservation	Water conservation forest	*Ph. edulis* *D. giganteus*
	Ecotourism	Scenic spot	*Ph. aurea* *B. ventricosa* *Thyrsostachys siamensis*

Table 2. Utilization of bamboo in China.

In terms of timber bamboo by roughing to finishing gradually, mainly using rough machining in building, not only can be used as scaffold material, scaffold and all simple bamboo house beam, column, wall, and bamboo ladder; deep processing of bamboo integrated from 80s, it began to gradually shift plywood, bamboo lattice plywood, bamboo particleboard, bamboo flooring and other, this is a new way for the utilization of bamboo.

The representative bamboo species of timber bamboo is *Phyllostachys edulis*. This species is generally known as Maozhu and is called Moso in the West. *P. edulis* is a major economic species grown in subtropical regions, which constitutes the largest artificial bamboo formation and has been thoroughly exploited and studied [12]. The topography varies from mountain-sides, rolling hills, and plains. This species is largely spread because of human cultivation. *P. edulis* covers a total area of 240 × 104 ha throughout China. Large forest areas are found in plains and mountain sides, approximately from 800 m above sea level to 1200 m in vertical distribution in some southward places(as shown in **Figure 1a**).

Another famous timber bamboo is *Dendrocalamus sinicus*. This species, also called "huge dragon bamboo" by Chinese locals, exhibits a unique feature of large thermal axis sympodial bamboo and is only located in the Southern and Southwestern Yunnan. *D. sinicus* can grow 30 m tall and 35 cm in diameter and is the world's biggest bamboo with great development value and utilization. The "huge dragon bamboo" provides excellent ecological protection and soil and water conservation because of its developed rhizome root system that reaches 5~10 m (as shown in **Figure 1b** and **c**).

Figure 1. The importance of bamboo species. (a) *Phyllostachys edulis*—bamboo forests—this Moso bamboo constitutes the largest artificial bamboo formation and has been thoroughly exploited and studied; (b) *Dendrocalamus sinicus bamboo*—it is the largest bamboo in the world, and it can be used to build houses; (c) bamboo sheath of *D. sinicus*—this sheath can be used to make hat and sole of shoe; (d) *Neosinocalamus affinis*—it is a very good weaving material; (e) culm bud of *N. affinis*—looks like Chinese word "山"; (f) *Fargesia fungosa*—*Fargesia* is an ecological and ornamental scenery species, and also is the main food source of panda. It also plays an important ecological role in soil and water conservation; (g) *Qiongzhuea tumidissinoda*—this bamboo has a high ornamental value and process value, and can be used to make cane. This species is distinguished for use in both art and crafts and for edible shoot production. Moreover, *C. tumidissinoda* is sold to other countries in South Asia since the Han and Tang dynasties and is listed in the first group of specially protected plants in China.

Bamboo shoots are used as traditional food by the Chinese people. The shoots can be used to make up for the nutrient deficiencies in the diet because of their nutritional and therapeutic values [13]. There are about more than 50 species of bamboo shoots used in China. Among them, the cultivation area is large, the number of shoots is high, and the bamboo shoots which are of good quality are available such as *P. edulis, P. dulcis, D. latiflorus, D. hamiltonii, Chimonobambusa quadrangularis*, and so on [14].

Bamboo can do some kind of woven material because of the long bamboo fiber and good tenacious, thus all kinds of bamboo can used for weaving. People use bamboo to weave basket, table, and chairs. *Neosinocalamus affinis* is the most widely cultivated bamboo, which is used to weave. Other bamboo varieties such as *Ph. nigra. var. henonis, Bambusa textilis, B. chungii*, and so on are also good weaving materials.

4. The traditional utilization

Bamboo has a tensile strength of 28,000 per square inch, which is higher than that (23,000 per square inch) of steel and is thus an essential material for structures that can withstand earthquake. Bamboo is also used to construct buildings. In remote southern mountains of China, some ethnic minorities live in bamboo houses. Bamboos are used as roofs and walls of houses, fences, domestic, and agricultural implements, such as water containers, food, and drink container hats, arrows, quiver, etc. The Chinese people eat bamboo shoots, which are a good source of dietary fiber and have low fat content and calories. Fresh, dried, canned, and flavored shoots are consumed as delicacies. In fact, bamboo shoot as food was used traditionally by tribal communities worldwide. To enhance the application of bamboo shoots, scholars have investigated them as food in Japan, Taiwan, Thailand, and other Asian countries and have developed several commercially available products in the market [15]. **Figure 2** shows some of the products made of bamboo.

5. Availability and potential use of bamboo resources

Bamboos are multipurpose plants of high economic and environmental value. Bamboo is one of the Non-Timber Forest Products (*NTFPs*) that has considerable potential as a wood substitute because of its high growth rate, good mechanical properties, and broad range of applications, especially in industrial fields. Bamboo has become a high-tech industrial raw material and substitute for wood.

According to data collected in this chapter, China contains the most abundant and diverse bamboo forest worldwide. A total of 861 species and infraspecific taxa belonging to 43 genera have been reported and include 707 species, 52 varieties, 98 forma, and 4 hybrids, which are naturally distributed in 21 provinces in China. The national bamboo forest area is 6.01 million ha, which include 4.43 million ha of Moso bamboo and 1.58 million ha of other bamboo species. The bamboo forest is mainly distributed in 18 provinces and mostly comprises Moso bamboo.

http://dx.doi.org/10.5772/intechopen.76485

Figure 2. Uses of bamboo: (a) Bamboo wine—bamboo wine is an original ecological food with zero contamination and is produced by injecting the finest sorghum wine into a young bamboo pole. After 3 years of brewing, the natural essence of wine is produced; (b) bamboo straw hat is made of woven bamboo leaves and wicker. The hat is sturdy and beautifully designed; (c) back basket—this backpack is a clothing worn by Huayao Dai minority endemic to Yunnan. This basket is worn at the back, can be decorated, and can hold small things. It has a role in beautification; (d) bamboo flute—the flute is a popular musical instrument in Chinese classical music and is made of natural bamboo; (e) bamboo pendant—Chinese words or pictures are engraved in these pendants for good luck; (f) bamboo shoot—bamboo shoot is used in Chinese traditional dishes and has a long history of eating and cultivation; and (g) bamboo charcoal—bamboo charcoal is widely used as fuel because it cannot be easily burned. Bamboo charcoal is also widely used in food, cooking, baking, storage, and preservation.

Although Yunnan has only 110.4 thousand ha of bamboo forest, it has the most diverse bamboo species (220 species), with many rare and endangered bamboo species. This province is one of the world's center of bamboo distribution. Bamboo forests can be divided into three types, namely cool temperate bamboo forests, warm and temperate bamboo forests, and hot bamboo forests. Different types of bamboos have various uses, commercial values, and ecological functions. Bamboo plays an immense role in the development of Chinese historical culture and the formation of Chinese ideology.

China has made a breakthrough in the development and scientific cultivation of bamboo. At present, China ranks first in bamboo research worldwide, because of its numerous bamboo research units and strong technical force. A bamboo research institute has been established in China Academy of Forestry Sciences, Southwest Forestry University, Nanjing Forestry University, Zhejiang Academy of Forestry, Jiangxi Academy of Forestry, and other forestry research institutions. In this chapter, we introduced the "International Bamboo and Rattan Center." The main duties and tasks of this center are as follows: to set up a national key laboratory for biological technology and material processing and utilization; to conserve bamboo and rattan; to build a world bamboo gene library; and to conduct international technology cooperation on bamboo resource protection, cultivation, material research, and development. In addition, many of the key bamboo timber-producing provinces and counties and large bamboo handicraft factories have set up a group of full-time staff engaged in bamboo science and technology work. Bamboo research covers a wide range of topics including the classification of germplasm resources, the application of bamboo, and the improvement in the aspects of high-yield cultivation, processing, and utilization [16].

Scholars must focus on discovering plants/crops that can be used as a new energy resource, particularly renewable energy. Energy development and utilization have become inevitable for sustainable development of economy and society. At present, China has limited energy supply, which is much lower than the world average supply of only 19.4% [17]. Bamboo is a renewable resource with many uses and may be considered as a lignocellulose substrate for bioethanol production because of its environmental benefits and high annual biomass yield [18]. Biological bamboo resources are mainly used in two ways. First, bamboos are used as raw materials in bioethanol production. Second, bamboos are used as raw materials for power generation [17]. The use of bamboos as new raw materials for energy supply in the bioenergy industry has opened up novel strategies and growth points for the development of the bamboo industry. Using bamboos to provide bioenergy is an important area to be explored in the future. This phenomenon requires careful and strategic planning based on adequate knowledge on the availability and potential use of bamboo resources for society and environment and on their production and growth behavior [19].

6. Conclusion

China has the most abundant bamboo resources worldwide and the richest bamboo culture. Bamboo greatly influences not only the Chinese history and culture but also every aspect of people's production and life, in clouding clothes, foodstuff, lodging, transport, and tools. In the future, the bamboo industry will be one of the pillar industries of the southern mountainous area's economic development; in addition, more than 80% of the bamboo forests in China are located at the source of river system, playing an important ecological function of soil and water conservation. Therefore, how to rationally protect and utilize the abundant bamboo resources in China will be a major problem to be solved in the future.

Acknowledgements

This work was financially supported by the 13th Five-Year Science and Technology Support Issues (2016YFD0600902, 2016YFD0600903), the National Scientific Foundation (31660173), the Key Project of Yunnan Provincial Education Department (515006009), and the Southwest Forestry University Talents Project 2012BAD23B04).

Author details

Weiyi Liu[1]*, Chaomao Hui[1], Fang Wang[1], Meng Wang[1] and Guanglu Liu[2]

*Address all correspondence to: weiyiliu651@126.com

1 Forestry College, Southwest Forestry University, Kunming, Yunnan, China

2 International Centre for Bamboo and Rattan Key Laboratory of Bamboo and Rattan, Beijing, China

References

[1] Mitch D, Harries KA, Sharma B. Characterization of splitting behavior of bamboo culms. Journal of Materials in Civil Engineering. 2010;**22**(11):1195-1199. DOI: 10.1061/(ASCE)MT.1943-5533.0000120

[2] Zhou G, Meng C, Jiang P, Xu Q. Review of carbon fixation in bamboo forests in China. Botanical Review. 2011;**77**(3):262-270. DOI: 10.1007/s12229-011-9082-z

[3] Pérez MR, Rodriguez LG, Yang X, Xie J, Fu M. From basic raw material goods to cultural and environmental services: The Chinese bamboo sophistication path. Ecology and Society. 2014;**19**(4):10. DOI: 10.5751/ES-06701-190403

[4] Fu JH. Chinese Moso bamboo: Its importance. Bamboo. The Magazine of The American Bamboo Society. 2000;**22**(5):5-7

[5] Guo QR, Yang GY, Du TZ, Shi JM. Carbon character of Chinese bamboo forest. World Bamboo & Rattan. 2005;**3**(3):25-28

[6] Chen HX. Promotion of bamboo for poverty alleviation and economic development. World Bamboo & Rattan. 2004;**2**(3):345-350

[7] Group BP. An updated tribal and subtribal classification of the bamboos (Poaceae: Bambusoideae). Bamboo Science & Culture. 2012;**24**(1):1-10

[8] Liese W. Bamboo—The gift of the gods. Journal of Bamboo & Rattan. 2003;**2**(3):297-298

[9] Yi TP. Yushania. In: Delectis Flora Reipublicae Popularis Sinicae Agendae Academiae Sinicae Edita. Beijing: Science Press; 1996. pp. 481-560

[10] Qiu E, Hong W, Zheng Y. Review on diversity and utilization of bamboo in China. Journal of Bamboo Research. 2001;**20**(2):11-14

[11] Hui CM, Hu JZ, Zhang GX. The status and prospects of bamboo diversity and the sustainable using in China. World Forestry Research. 2004;**17**(1):50-54

[12] Hu MF. Status of diversity conservation and using of bamboo plants in Fujian. Bamboo Research. 2002;**21**(4):33-38

[13] Devi YR. Bamboo forest resources of India and its role in food security: a review. Agricultural Reviews. 2013;**34**(3):236. DOI: 10.5958/j.0976-0741.34.3.009

[14] Zhou FC. Bamboo Stands Cultivation. Beijing: China Forestry Press; 1998. p. 17

[15] Bal LM, Singhal P, Satya S, Naik SN, Abhijit K. Bamboo shoot preservation for enhancing its business potential and local economy: A review. Critical Reviews in Food Science and Nutrition. 2012;**52**(9):804-814. DOI: 10.1080/10408398.2010.511321

[16] Huang T. Exploitation and utilization of bamboo forest resources in China. World Agriculture. 1998;**6**:46-47

[17] He MX, Wang JL, Han Q, Shui ZX, Zhu QL, Bo W, et al. Bamboo: A new source of carbohydrate for biorefinery. Carbohydrate Polymers. 2014;**111**(20):645-654. DOI: 10.1016/j.carbpol.2014.05.025

[18] Yao WL, Xin T. Analysis of energy situation and development countermeasures in Ch. Energy research and Information. 2006;**22**(4):187-193

[19] Wu ZZ, Xia EL, Wang SD, Zhong ZK. Development and utilization of bamboo bio-energy and its prospects in China. World Forestry Research. 2013;**26**(2):60-64. DOI: 10.13348/j.cnki.sjlyyj.2013.02.006

Section 4

Potential and Challenges

Chapter 9

Recent Advancement in Physico-Mechanical and Thermal Studies of Bamboo and Its Fibers

Samsul Rizal, Abdul Khalil H.P.S., Ikramullah, Irshad Ul Haq Bhat, Syifaul Huzni, Sulaiman Thalib, Asniza Mustapha and Chaturbhuj Kumar Saurabh

Additional information is available at the end of the chapter

http://dx.doi.org/10.5772/intechopen.76475

Abstract

Bamboo has its own role in the development of society. It is widely used as a support tools for sustainable farming and being exceptional resource for variety of income and employment-generating systems. This overlooked biomass provides food, raw material, shelter, medicine for large part of world's population. Bamboo has given a great support to mankind by providing building materials that are extensively used for household products. It has found a good place for industrial applications due to advances in processing technology and increased market demand. Numerous fundamental studies were carried out to highlight their fundamental characteristics prior to industrial exploitation or high end bamboo-based biomaterials. The mechanical and physical properties of bamboo have noteworthy effects on their durability and strength. Thermogravimetry is one of the key sensitive technique that characterizes the mechanical responses of materials by monitoring property changes with respect to the temperature. Comprehensive review and study on thermal analysis are useful for quantitative determination of the degradation behaviour and the composition of the material. The magnitude and location of the derivative thermogravimetric curve also provides information of the interaction between the material components at various temperature scales. Therefore, these studies can be focused to evaluate the basic fundamental problems faced and thus, a well-designed research and development towards sustainability can be achieved.

Keywords: physico-mechanical, thermal properties, thermogravimetry, bamboo

1. Introduction and global scenario

Bamboos are an extremely multipurpose and most beneficial group of plants. Bamboos are found in almost all continents of the world. Both common people and researchers have utilized bamboo in a variety of ways and have played an important role in today's life. At present, bamboos are considered as highly renewable resources [1, 2]. The development and exploration of bamboo eventually depends, not only on scientific community, but on urban and rural people who are largest utilizers (buyers, sellers and processors) of bamboo. Bamboo has played an important role in the lives of people in South and Southeast Asia [3]. Being an essential natural resource, "sustainable development" is a critical concern of the bamboo economy. The development of bamboo is a concern of both enhancing the supply and ensuring productivity, which leads to development and needs of present and future. Presently non-profit organizations like American Bamboo Society, International Network for Bamboo and Rattan (INBAR) Latin America, INBAR West Africa, European Bamboo Society, INBAR East Asia, INBAR West Asia and Bamboo Society of Australia have clear objectives to promote all aspects of bamboo interest in respective countries [4].

These organizations promote opportunities for sharing information and education, and awareness among common people. The role of bamboo in plantations and sustainably have achieved forests in providing economically possible environmental services. Different resource inventory methods and technical manuals have been considered by these organizations to improve criteria and indicators for sustainable resource management along with income generation for local people, particularly for bottom billion people. In addition to these societies, researchers are contributing to make a connection of sustainability with the world growing population, average welfare rate and environmental impact of welfare commodities, demonstrating the need of achieving a factor 20 environmental improvement by the year 2040. The material scientists and engineers [5–8] can take up the responsibility to contribute to sustainable development by finding more environmentally benign ways of research in material sciences and engineering applications. One of the ways for solutions is to be found in new material applications: improving the dimensional stability of bamboo products, providing practical methods and procedures in determining the strength and thermal stability of bamboo and their products.

The use of bamboo can be explored for various purposes depending upon its properties. It plays an important role for house construction, agricultural tools and implements, as food material, weaponry, bicycles, etc. [9, 10]. The traditional cottage industry has been greatly benefited and it is also an appropriate source of cellulose for paper production and rayon [11]. Bamboo craft is one of the known of traditional cottage industries. Since early ages the baskets, mats and many other products of household use were fabricated from bamboo as it is available in vicinity of nearby forests. Later, tribal and rural people took up bamboo as a means of livelihood. Now bamboo craft is well established in all rural areas of the country and it feeds millions of traditional workers [12].

Bamboo is mainly grown in tropical regions of Asia, Latin America and Africa. Bamboo can be called as a collective name for different species of giant grasses [13]. It has been estimated

Use of bamboo as plant	Use of bamboo as material
Ornamental horticulture	**Local industries**
Ecology	Artisanat
Stabilize of the soil	Furniture
Uses on marginal land	A variety of utensils
Hedges and screens	Houses
Minimal land use	**Wood and paper industry**
Agroforestry	Strand boards
Natural stand	Medium density fiberboards
Plantation	Laminated lumber
Mixed agro-forestry system	Paper and rayon
	Parquet
	Nutritional industries
	Young shoots for human consumption
	Fodder
	Chemical Industries
	Biochemical products
	Pharmaceutical industry
	Energy
	Charcoal
	Pyrolysis
	Gasification

Table 1. Various uses of bamboo [16].

that 60–90 genera of these species are available in different sizes and forms. Nature has given bamboo special structural design. Due to the direction of fibers and hollowness of bamboo less material is needed as compared to materials with high mass content e.g. wood. This unique property has led a world class revolution in bamboo both in cultivation and utilization. A comprehensive range of bamboo products are available in variety of styles and design patterns in the form of furniture, decorative products, wall hangings, basket, lamp shade, and table top items. The mostly rural people have used bamboo crafting as a tool for designing such products. In comparison to ancient era where it has been used as household purposes for keeping goods, the modern scenario is different as the demand of bamboo is persistently growing in the domestic as well as international markets due to its varied design patterns as well as its eco-friendly features.

It has been estimated that at global level about 2.5 billion people use bamboo in different forms with an annual turnover of 10 billion USD and the figures will be doubled by 2015. China is the largest among the bamboo exporting countries. The total export value of bamboo products of the country is $550 billion per annum [14]. Moreover, advanced research activities have been carried out for efficient fuel generating systems. The international network for bamboo and Rattan (INBAR) are functioning at the international level. A well mechanized mode has been adopted at global level, for primary processing and product manufacturing. A collective effort by Choudhury and Sharma [15] to summarize the uses and importance of bamboo shoots on global level has provided enough knowledge to researchers associated with development and utilization of bamboo. The authors provided a global scenario of marketing the bamboo and earnings from it although considering it as underestimated natural resources at international level, even being found naturally or cultivated. The market value of bamboo in China and America is worth of millions dollars in terms of import and export marketing. USA imports tones of bamboo from many Asian countries for food items. India, an economically superpower country is alone doing bamboo marketing in millions of rupees. As summary, various uses of bamboo are presented in **Table 1**.

2. Role of bamboo in sustainable economic viability

World bamboo sector in different countries are unique in terms of their scale and level of organization. Bamboo in many countries follows the global constraints on forests and timber, and the ongoing need to develop rural industries and improve poverty. Policies in the forest sector are critical in ensuring the healthy development of the bamboo. Depending on the particular condition, bamboo is treated as tree or agriculture crop. To avoid hindrance to sustainable development, this tree or crop is fitted in such policies to pose least threat to sustainable sector. Therefore, clear classification had been designed by different countries, which are well recognized and made in accordance with ecological conditions of bamboo. Bamboo has been recognized as multifunction species and single purpose for bamboo has been avoided by developing countries. In economically sound countries with significant bamboo sectors, different agencies and societies are working specifically on bamboo are playing an important role in formulating policies to ensure healthiness of agro-forestry and natural resources. For example, model systems opted by China should be utilized where there is strict need of it. Presently, the farmers are investing in the bamboo industry with making secure their right of uses manage and transfer resources.

The restricted use of bamboo in some contents of the world has provided limited knowledge about the economic feasibility of bamboo. To acquire more knowledge into building and maintenance costs of the initially described bamboo applications have been compared to more common building materials such as steel, timber and concrete taking environmental assessment and the same structural elements into account [17]. Some buildings have been executed in European countries with bamboo as the main structural material. Although, there are problems some of which had direct consequences of the use of bamboo. This influence of working with bamboo has been analyzed. Therefore, studies on critical factors of failure and success of the application of bamboo in Western European building projects have been carried out to provide solutions as to preventing measures or reduce the negative impacts of their causes [17].

http://dx.doi.org/10.5772/intechopen.76475

No.	Species	Local name	Uses
1	*Bambusa blumeana*	Buluh duri	Chopstick, tooth picks, furniture, musical instrument, poles, shoot as food
2	*Bambusa heterostachya*	Buluh galah/tilan/pering	Poles, frames, tooth picks, blinds, skewer sticks
3	*Bambusa vulgaris*	Buluh minyak/aao/ aro/gading/ tamalang	Ornamental, tooth picks, chopsticks, skewer sticks, shoot as food
4	*Bambusa vulgaris var. striata*	Buluh gading	Ornamental
5	*Dendrocalamus asper*	Buluh betong/pering	Shoots as food, chopsticks
6	*Dendrocalamus pendulus*	Buluh akar/belalai	Handicraft, basket
7	*Gigantochloa brang*	Buluh brang	Shoots as food, chopsticks, skewer sticks, tooth picks
8	*Gigantochloa levis*	Buluh beting/bias	Shoots as food, chopsticks
9	*Gigantochloa ligulata*	Buluh tumpat/tikus belalai	Frames, shoots as food, poles for vegetable support
10	*Gigantochloa scortechinii*	Buluh semantan	Handicraft, small scale industries, incense sticks
11	*Gigantochloa wrayi*	Buluh beti/raga	Handicraft, small scale industries, incense sticks
12	*Schizostachyum brachycladum*	Buluh nipis/lemang	Handicraft, rice vessels (lemang)
13	*Schizostachyum grande*	Buluh semeliang/ semenyeh	Frames, leaves used for wrapping Chinese glutinous rice dumpling
14	*Schizostachyum zollingeri*	Buluh dinding/kasap/ telor/nipis	Handicraft, tooth picks, skewer stick
15	*Bambusa multiplex*	—	Fishing rods
16	*Bambusa tuldoides*	—	Ornamental
17	*Gigantochloa balui*	—	Handicrafts, sailing masts
18	*Schizostachyum pilosum*	—	Flooring and basketry
19	*Dinochloa darvelana*	—	Pests species
20	*Dinochloa obclavata*	—	Pests species
21	*Dinochloa prunifera*	—	Pests species
22	*Dinochloa robusta*	—	Pests species
23	*Dinochloa scabrida*	—	Pests species
24	*Dinochloa sipitangensis*	—	Pests species
25	*Dinochloa sublaevigata*	—	Pests species
26	*Dinochloa trichogona*	—	Pests species

Table 2. Commonly available bamboo species in Malaysia [20].

In Malaysia, government has classified bamboo as a non-timber forest and is next in importance to rattan. It has been used as food as well as traditional and commercial products since ages. There are more than 50 species of bamboo reported in Malaysia, an integral part of forestry, but it is also commonly spread outside forests including farmlands, riverbanks, roadsides and urban areas. Based on Fourth National Forest Inventory, bamboo occupies about 7% of the total forest area in Peninsular Malaysia (the total area of Peninsular Malaysia is 131,600 km^2). The list of bamboo species available in Malaysia with proper uses is given **Table 2**. The genera found in the country are *Bambusa, Dendrocalamus, Dinochloa, Gigantochloa, Racemobambos, Schizostachyum, Thyrsostachys, Chusqua, Phyllostachys* and *Yushania* [18]. Among all the species available, 14 Malaysian bamboo have been identified as commercial species [19] To maximize the exploration of bamboo industry in Malaysia, bamboo resources are needed to be maintained for industries particularly for joss sticks, chopsticks, basket-making, toothpicks and joss-papers. A systematic policies for the production bamboo is the need of present economic scenario of Malaysia, otherwise there will be depletion in demand of bamboo. The sustainable future of bamboo in Malaysia needs sustainable plantation of bamboo resources. In context to this, Malaysian government has disseminated trial of Malaysian commercial bamboos at few sites but not on the whole Peninsular Malaysia.

3. Physico-mechanical properties of bamboo for commercial utilization

The advancement of science and technology has led to new methods to make bamboo more durable and usable in terms of building materials. Bamboo has helped to uplift the rural economy through the establishment of industries and opportunities of employment. Bamboo based industries could be a sustainable option to ease the domestic demand and bring foreign currency by exporting newly designed products to international markets. The use of bamboo can be exploited to greater extent as it is cheap and found in abundance. The properties like rapid growth rate, short rotation age, excellent flexibility and high tensile strength has transformed bamboo into a wide variety of products ranging from domestic household products to industrial applications [21]. The uses in platforms (floors for transport vehicles such as trucks, busses and rail coaches), concrete molding boards (in building industries), flooring, furniture, pulp and handicraft works has provided new dimensions to bamboo utilization [22–24]. To elevate the utilization of bamboo, its fundamental physical and mechanical properties must be fully understood [25]. Researchers have proved that physical and mechanical properties of bamboo vary with respect to position in the bamboo. Lee et al. [26] found that the physical and mechanical properties of bamboo are affected by height location. Xian and Ye [27] studied the variation in mechanical properties of bamboo and established an equation for predicting the tensile modulus of elasticity from the radial position. Li [28] investigated the variation of the specific gravity and bending properties of bamboo and found that the specific gravity and bending properties decrease from the outer to inner layers of the bamboo culms.

3.1. Physical properties

3.1.1. Moisture content

Utilization of bamboo has now advanced from traditional to structural applications such as composites and advanced materials [29, 30]. The advancement in usage of bamboo needs further understanding of the material characteristics such as the physical properties. Terminology of a bamboo culm is illustrated in **Figure 1**. Physical properties of the node and internode positions of bamboo have been investigated by Tamizi [31] with small size specimens (strips of bamboo). The statistical data obtained showed a great variation according to the sources and position of the samples obtained from the bamboo. It was observed that moisture content was higher at the inner layer and reduced in the outer layer of the bamboo culm. Liese [32] claimed that different bamboo species showed different moisture values which can be attributed to difference in some inherent factors such as age, anatomical features and chemical composition. But in this case, the age factor is not involved since all samples were taken from 3 years bamboo culms. In this chapter, discussion is focusing on moisture content, specific gravity, shrinkage and fracture roughness.

The higher moisture content could be influenced by the anatomical structure of bamboo. The inner layer contains lower vascular bundles concentration which leads to higher moisture content as compared to outer layer as shown by Li [28]. This phenomenon is similar to non-wood plant, i.e., oil palm trunk which shown higher content of parenchyma in core part. Engler et al. [34] has reported the relation between moisture content and thermal use of one of the bamboo species. Authors stated that moisture content is one of the most relevant characteristics, which significantly influences the thermal use and efficiency. A comparison study with other species of wood has been done and authors revealed that the moisture content of bamboo was higher at an average of 136.9% and spreading widely. Related to the ages, the

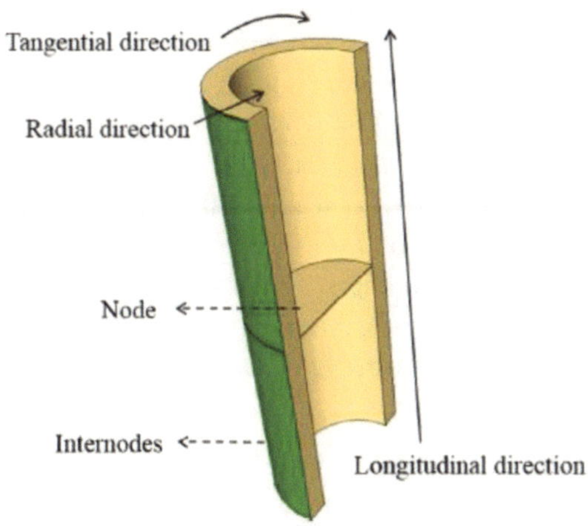

Figure 1. Terminology of a bamboo culm [33].

young culms in general show significant higher moisture contents, compared to the older culms. It was also found that moisture content at the bottom of bamboo culm was higher as compared to the top.

Studies by Kamthai [35] on different physical and mechanical properties of sweet bamboo found that the moisture content was 60.2%. On the other hand, Chen et al. [36] investigated the moisture content of modified bamboo strips. Alkaline treatment enhanced the moisture absorption, while esterification treatment, oxidation and silane treatments has reduced the moisture content. The results revealed that moisture content directly affects the other properties like interfacial shear strength.

3.1.2. *Specific gravity*

Different observation by various researchers has been put forward in order to get in-depth knowledge about the specific gravity of bamboo, for example a study on nodes and internodes of *Gigantochloa* has been carried out by Tamizi [31]. The specific gravity of samples from outer layer was higher compared to middle and inner layer for both internodes and nodes of all the *Gigantochloa* species. *G. levis* node recorded the highest specific gravity value among all of the species while *G. wrayi* node gave the least values. The bamboo density has a close relation with vascular and ground tissues percentages as proposed by Janssen [37] and Espiloy [38]. It was revealed that specific gravity of internode and node part of each bamboo species was marginally different. In contrast to report by Hamdan et al. [39], it was found that the nodes present along the culms height generally have higher density than the internodes due to lesser presence of parenchyma as well as lower moisture content and volumetric shrinkage.

3.1.3. *Shrinkage and fracture toughness*

Shrinkage is a characteristic property of bamboo which describes tendency of bamboo towards shrinkage under different conditions. Unlike wood, bamboo has a tendency to shrink from the very beginning of drying. The elimination of moisture in the cell wall that is hygroscopic or bound water leads to shrinkage as a result of the contraction of microfibrillar net in proportion to the amount of liquid evaporated [40]. Yu et al. [22] reported a study on shrinkage at different locations of Moso bamboo (*Phyllostachys pubescens*). The results revealed that both height and layer had a substantial effects on tangential and longitudinal shrinkages, but the interaction between height and layer had no significant effect on shrinkage. It was observed that tangential and longitudinal shrinkages appeared to be divided into two 3-layer zones (i.e., outer 3-layers consisting of layers 4, 5 and 6 and inner 3-layers consisting of layers 1, 2 and 3). It was found that tangential shrinkage was slightly greater at 4.0 m and longitudinal shrinkage was slightly greater at 1.3 m.

Amda and Untao [41] studies about the physical properties such as fracture toughness and tensile tests of bamboo culm and nodes. Authors reported that fracture toughness of bamboo culms depends on the volume fraction of fibers. It was observed that bamboo culm has a high value of fracture toughness for outer surface layer and decreases towards the inner surface, meaning that bamboo offers a greater fracture toughness on the outer surface where the most

http://dx.doi.org/10.5772/intechopen.76475

dangerous and external force is exerted. Authors also explained that fracture toughness distribution with radius was in accordance to the values obtained for volume fraction of fibers, which means the proportionality depends directly to each other. The value of fracture toughness is increased at outer surface while moving towards outer surface. Moreover, it was found that fracture toughness of bamboo was higher than aluminum alloy and way better than other wood species.

3.2. Mechanical properties

Bamboo is known for its orthotropic character meaning, it has specific mechanical properties in x, y, z directions i.e., longitudinal, radial, and tangential. A pool of knowledge of the mechanical properties of bamboo helps in safe design as bamboo responses in the same manner as other building materials do. However, being a natural composite or biological material like wood, it is exposed to greater variability and complexity, due to various growing conditions as availability of moisture and soil conditions. Preparation of bamboo samples prior to mechanical testing is presented in **Figure 2**.

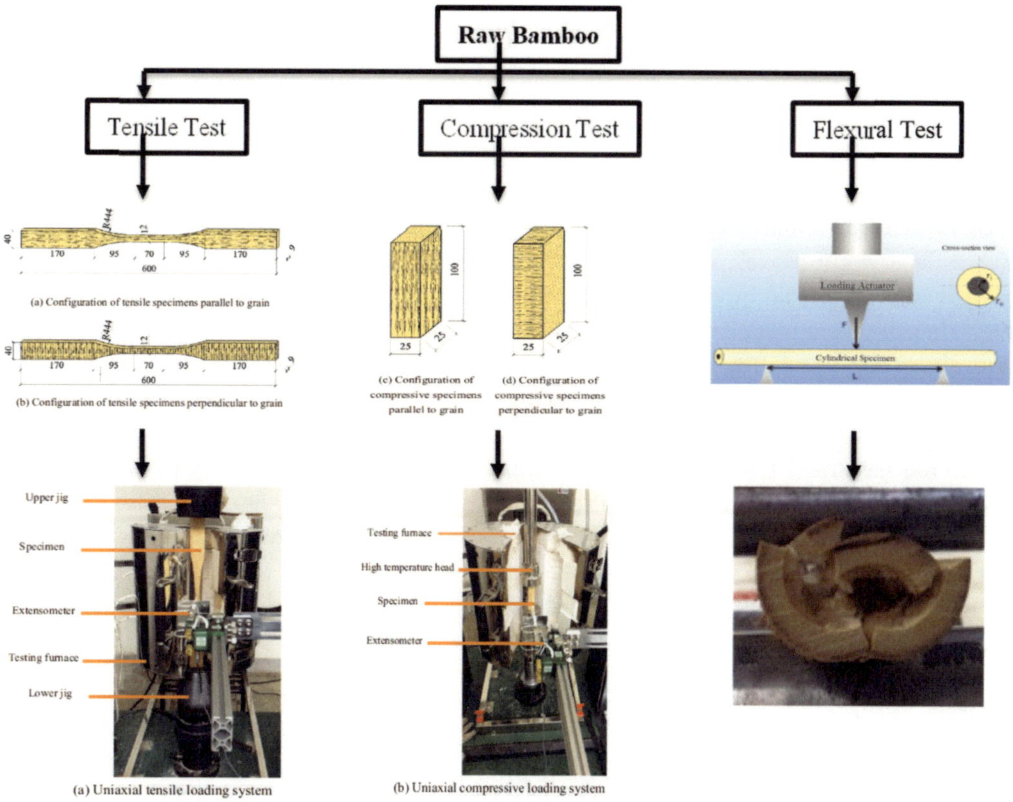

Figure 2. Preparation of bamboo samples for mechanical testing including tensile test, compression test and flexural test [9, 42].

3.2.1. Tensile strength and modulus

Studies have been carried out to investigate the variation of mechanical properties as well as between the internodes and nodes, and the variation between different locations in the bamboo culm [43, 44]. Researchers have conducted studies on the mechanical behavior of both full size culm (round form) [45, 46] and small specimens [47, 48]. Lakkad and Patel [49] revealed that the specific strength of bamboo is greater than that of the most thermosetting resins. It was observed that bamboo is stiffer and stronger than other woods. Therefore, authors hypothesized that bamboo has remarkable potential as a structural material when considering the mechanical properties of bamboo together with its low cost properties.

Defoirdt et al. [50] assessed the tensile properties of bamboo fibers to understand, how they go well as starting material for composite material. They reported that decreasing trend in strength of bamboo fiber was observed as length was increased justifying that, with higher test lengths there are more drawbacks in the fibers that increases the chances of failure. They compared the tensile strength of bamboo with other fibers and found that tensile strength of bamboo is higher in contrast to other fibers even though the bamboo fibers were damaged during extraction process. In other study, Tan et al. [51] revealed that the tensile strength degradation of bamboo corresponds to the fiber density degradation. Authors reported the highest strengths were found to be in the regions nearer outer surfaces with higher fiber densities. However, lowest strengths correspond to the regions away from the outside surfaces with the lowest fiber densities whereas in between regions shown intermediate tensile strengths.

Li [28] also stated that tensile strength and mean Young's modulus increases with increase of cellulose content and decreasing microfibrillar angle. Rao and Rao [52] carried out a study on tensile strength as well as tensile modulus of bamboo. They reported that bamboo fibers possess highest tensile modulus as compared to other fibers viz., palm coconut and sisal fibers. In our recent work carried out on *Gigantochloa sps*, we observed that the higher tensile modulus group was *G. levis* (3793 MPa), followed by *G. wrayi* (3670 MPa) and *G. scortechinii* (3456 MPa) and the lower was *G. brang* (2661 MPa); (*G. levis* > *G. wrayi* > *G. scortechinii* > *G. brang*). There was a noteworthy difference between green and air dry sample. This might due to the fact that bamboo behaves as similar to wood whereby the mechanical properties increases with the decrease in moisture content [39]. The analysis of variance for tensile modulus at different locations exhibited a significant difference between the internode strips and node strips.

3.2.2. Static bending (modulus of rupture and modulus of elasticity)

Modulus of rupture is the maximum flexural stress sustained by the specimen during bending test or in other words can be defined as the maximum stress in bending that can be withstood by the outer fibers of a specimen before rupturing. Tamizi [31] reported the comparative results of modulus of rupture with reference to different position of different species of bamboo. Both dry and green sample exhibited different values of modulus of rupture. The order was (*G. scortechinii* > *G. wrayi* > *G. levis* > *G. brang*). It was reported that air dry samples showed better modulus of rupture (30–40%) compared to green samples. This can be ascribed to the fact that bamboo behaves similar to wood whereby the mechanical properties increases with decrease in moisture content. The data presented that results at inner surface (130.71 MPa) direction were higher compared to the outer surface direction (111.07 MPa). The difference in MOR showed that the outer layer has less than 5% stronger from the middle layer, and the middle layer has less than 5% stronger than inner layer. Difference in MOR, the outer layer strength was less than 10% higher than the inner layer. The value for MOR (green) at the node is relatively higher than the internode. This is reflected due to the density of the node are higher compared to the internode as node contain lesser presence of parenchyma. This is similar to the earliest report mentioning that the strength properties in bending of bamboo are greatly correlated with specific gravity [53].

Biswas et al. [54] reported utilized wastes from two species of bamboo viz. *B. balcooa* and *B. vulgaris*. The waste was a type of shavings acquired during planning operation of bamboo splits for production of rectangular strips of constant thickness. This raw material was utilized for the formation of particle board. The modulus of rupture (MOR) values of boards made from planer waste was 15.7 N/mm^2 and 17.7 N/mm^2 for *B. balcooa and B. vulgaris*, respectively. Varied results were obtained and particle geometry had significant influence on modulus of rupture in both species. The bamboo planer waste had no definite pattern of particle geometry with light and curly shape with good tendency to fold. Authors revealed that based on modulus of elasticity a significant effect was observed on the particle board geometry.

Based on modulus of elasticity Li et al. [55] published an interested report that bamboo has high modulus of elasticity as compared to human bones. The average modulus of elasticity across the thickness of bamboo culm can be 18 GPa, equivalent to that of human cortical bone. Ghavami et al. [56] studied the modulus of elasticity of bamboo with some degree of precision. Authors stated that fiber distribution in bamboo follows an organized pattern with an increased amount of fibers on the outer surface of the culm. While proving how this variation occurs, the basic equations from the bamboo approach can be changed in order to model the mechanical behavior of bamboo. Yu et al. [22] revealed that modulus of elasticity of bamboo varied greatly from the inner layer outwards. Modulus of elasticity at 1.3 m was less than those at 4.0 m from the base. This report suggests that layer had important effect on the modulus of elasticity. It was observed that modulus of elasticity decreased as the layer decreased from the outer (layer 6) to the inner (layer 1) layers, but the difference between layers 3 and 4, and between layers 1 and 2 were not considerably less.

4. Thermogravimetric studies of bamboo

4.1. Thermogravimetric analysis (TGA) and derivative thermogravimetric (DTG)

Thermogravimetric analysis (TGA) is a useful method for the quantitative determination of the degradation behaviour and the composition of particular material. The magnitude and location of the curve in thermograms provides the information of the component and the interaction between the components at various temperature scales [57]. The chemistry of bamboo is complex and has been divided by analytical methods into major components including cellulose, hemicellulose and lignin like several other lignocellulosic materials [58, 59]. Under these circumstances, the thermal decomposition is somewhat similar to the combination of those of the constituents pyrolyzed at the same conditions.

In a study by Tamizi [31] on different species of bamboo (*Gigantochloa*), viz., *G. brang* and *G. levis*, TGA studies were carried out at different positions which designated as outer, middle and inner layer. The shape of thermograms in these two species revealed quite similar characteristics. The behavior was in agreement with the data obtained by other researchers. Xie et al. [60] and Mui et al. [61] reported that the temperature range, weight losses and the rates of thermal degradation at different stages (devolatilization and combustion steps) changes with each different fiber at any specific location of the plant [62]. Thermal stability was observed up to 210°C in *G. brang* and thereafter started to decompose.

The initial degradation temperature between 210 and 390°C are linked to decomposition of bamboo constituents, which are mainly cellulose and hemicellulose [63, 64]. The second stage degradation between 390 and 800°C in the entire sample corresponds to decomposition of lignin. Previous studies have shown that the thermal stability is determined by the chemical composition of any biomass as various components of lignocellulosic materials have different thermal behaviors [62, 65]. Also, several studies on the thermal decomposition of the individual components of lignocellulosic materials indicated that decomposition of hemicellulose starts first, followed by the cellulose and finally, the lignin [65–68]. Similarly, in the thermograms of *G. levis*, samples gave an initial weight loss below 100°C attributed to the loss of moisture. Thereafter, the thermal degradation of the bamboo samples starts to decompose near 200°C at both internode and nodes. A distinct weight loss was observed between 230–400 and 230–390°C at the internode and node, respectively.

Differential thermograms revealed that different mass losses due to different constituents present at different positions. From DTG curves, it can be concluded that two major processes took place when *G. brang* decomposed. It was observed that a minor weight loss occurred below 100°C, with peaks between temperatures of 25 and 105°C. These weight losses have been reported to be associated loss of water as a moisture evaporation process [65]. The main DTG peak was assigned to the decomposition of hemicellulose and cellulose [69]. The decomposition of lignin is indicated by clear wide tail due to devolatilization process [63]. In presence of nitrogen, it has been observed that mass loss of small biomass samples at lower heating rates usually produce one to two major distinct DTG peaks, corresponding to hemicellulose and cellulose pyrolysis. For example, Font et al. [70] inspected the thermal decomposition of almond shells at 10°C min^{-1} in the inert atmosphere, resulting in two not completely separate DTG peaks, with one centering at around 310°C and other at 368°C.

In another study, one major peak was recorded for degradation of cotton stalk, sugarcane bagasse, rice straw, EFB, etc., under nitrogen while two peaks were reported by the same author for the degradation of these bioresources under oxygen. Thus, changes in the parameters like heating rate and atmospheric condition can sometimes merge the two peaks into one very broad peak [70, 71]. Other study has reported thermogravimetric study on alkaline treated *Bamboosa balcua* both in strip and dust form. They observed that thermograms of untreated bamboo exhibited two steps of degradation in wide range between 50–150 and 426–150°C, respectively. The first step of degradation, 7.65% weight loss was assigned to moisture evaporation and variation in trend in weight loss was observed in treated samples depending upon concentration of alkaline used. It was found that with increase in concentration of alkaline the amount of moisture absorbed was decreased, which was evident from TGA curves. This observation was supported on the basis of crystallinity index [72]. The tendency to release absorbed moisture upon heating will decrease as water is strongly attached in a well packed structure leading to higher finished temperature.

Typical TGA curves determined and reported by researchers can be illustrated by **Figure 3**. Apart from this, the thermal degradation process of bamboo fiber reinforcement was studied by Rajulu et al. [73] and it was stated that the thermal degradation of the fibers follow a two stage process. The TGA curves obtained might due to the fact that bamboo is composed of a strong composite network, which is interlinked through inter and intramolecular hydrogen bonding between polyphenolic groups lignin, hemicelluloses and α-cellulosic components of this network. The second degradation temperature region corresponds to hemicellulose degradation present in bamboo fiber.

A study on different hemicellulosic subfractions of *Phyllostachys bambusoides* by Peng et al. [76] to explore the mass loss and compare thermal behavior of different fraction was carried out. It was observed that weight loss of among two isolated fractions (HA and H_{45}) mainly occurred in the range of 200–320°C and was found that HA fraction was more thermally stable than H_{45}, indicating that hemicelluloses exhibited more thermal stability than branched hemicelluloses. This observation was supported by DTG studies, where the DTG curves exhibited

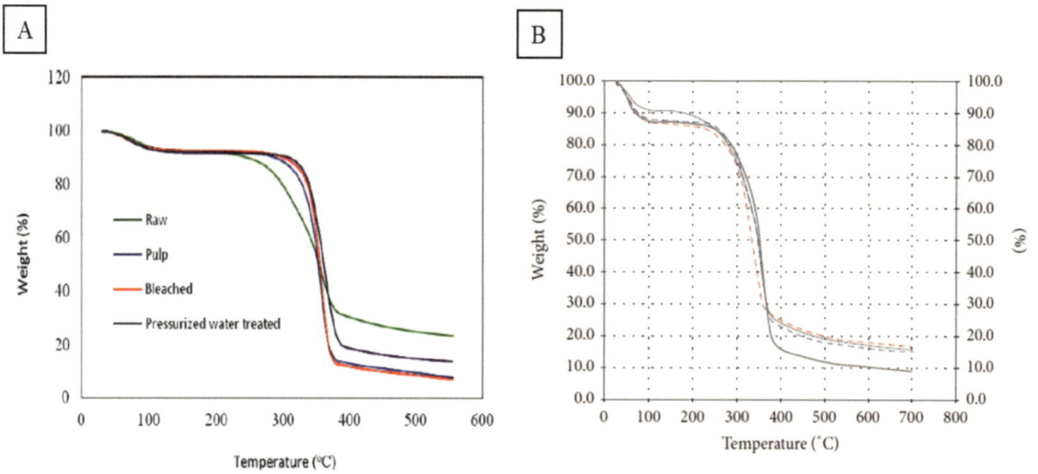

Figure 3. TGA thermograms of (A) raw, pulp, bleached, and pressurized enzyme hydrolysis bamboo fibers [74]; and (B) thermal analysis of green bamboo fiber, dewaxed bamboo fiber, delignified bamboo fiber and cellulose fiber [75].

the maximum peak at 296 and 293°C with shoulders at 242 and 237°C, respectively for H_{45} fraction. These results were justified by different researchers with different view of explanation. Nowakowski et al. [77] corroborated the first peak with reduced temperature polymerization process leading to the formation of char, carbon monoxide, carbon dioxide and water. Meanwhile, the second peak was assigned to generation of volatile anhydrosugars and related monomeric compounds.

Mui et al. [61] reported that three key components of bamboo namely, xylan, cellulose, and lignin has one major decomposition step to volatiles and a minor decomposition to which leads to the formation of char. A well model theory was presented to explain the decomposition of these components from bamboo to volatiles and chars called as five component and six component systems. Different thermo-parameters were obtained by these model systems. Moreover, Krzesinska et al. [78] conducted the thermal studies on solid iron bamboo (*Dendrocalamus strictus*), a bamboo with unique properties. The thermal decomposition of *Dendrocalamus strictus* was studied at different temperatures varying from 300 to 600°C. It was found that in the case of both raw and pre-charred *Dendrocalamus strictus,* poor thermal decomposition is completely absent. These carbonized samples of *Dendrocalamus strictus* would be a bonus for manufacturing of thermally stable composites.

5. Conclusion

The foregoing discussions have explored the broader aspects of bamboo, cast as it were within the discourse on sustainable development. To conclude, the above findings of pertinence are worth reiterating here in debating the future challenges confronting the bamboo industry, even though the bamboo industry has contributed a lot to Malaysia's economy. Land use policy choice is becoming more concerned as conservation groups have asserted that much of the land is being cleared to grow bamboo, while increasing the carbon dioxide content of the atmosphere, bamboo is the best plant to utilize this carbon dioxide. There is still considerable scope for increased utilization and value added products from bamboo. Malaysia stands to benefit from such a policy turn as bamboo industry is able to participate in cutting edge agricultural research and innovation. To get in-depth knowledge about the structural use of bamboo the study on fundamental properties is a necessary step towards the effective utilization of bamboo in market. With more research, a large number of value-added bamboo products for higher profitability could be made available in the market. If implemented effectively, the policy turn to diversify bamboo products will reflect the bamboo industry's green revolution effort towards a sustainable future.

Acknowledgements

The authors would like to thank Syiah Kuala University, Banda Aceh, Indonesia, and Universiti Sains Malaysia (USM) for the research collaboration.

http://dx.doi.org/10.5772/intechopen.76475

Author details

Samsul Rizal[1]*, Abdul Khalil H.P.S.[2], Ikramullah[1], Irshad Ul Haq Bhat[3], Syifaul Huzni[1], Sulaiman Thalib[1], Asniza Mustapha[2] and Chaturbhuj Kumar Saurabh[2]

*Address all correspondence to: samsul.rizal@unsyiah.ac.id

1 Department of Mechanical Engineering, Syiah Kuala University, Banda Aceh, Indonesia

2 School of Industrial Technology, Universiti Sains Malaysia, Penang, Malaysia

3 Faculty of Earth Science, Universiti Malaysia Kelantan, Campus Jeli, Kelantan, Malaysia

References

[1] Li Y, Yao J, Li R, Zhang Z, Zhang J. Thermal and energy performance of a steel-bamboo composite wall structure. Energy and Buildings. 2017;**156**:225-237

[2] Liese W, Kohl M. Bamboo. New York City, United States: Springer International Publishing; 2015

[3] Huang Z, Sun Y, Musso F. Assessment of bamboo application in building envelope by comparison with reference timber. Construction and Building Materials. 2017;**156**:844-860

[4] Ling M, Christensen M, Donnison A, Belmonte KD, Brown C. Scoping Study to Inform the Global Assessment of Bamboo and Rattan (GABAR). Cambridge, United Kingdom: United Nations Environment Programme (UNEP); 2016

[5] Sánchez ML, Morales LY, Caicedo JD. Physical and mechanical properties of agglomerated panels made from bamboo fiber and vegetable resin. Construction and Building Materials. 2017;**156**:330-339

[6] Li M, Zhou S, Guo X. Effects of alkali-treated bamboo fibers on the morphology and mechanical properties of oil well cement. Construction and Building Materials. 2017;**150**: 619-625

[7] Sukmawan R, Takagi H, Nakagaito AN. Strength evaluation of cross-ply green composite laminates reinforced by bamboo fiber. Composites Part B: Engineering. 2016;**84**:9-16

[8] Bahari SA, Krause A. Utilizing Malaysian bamboo for use in thermoplastic composites. Journal of Cleaner Production. 2016;**110**:16-24

[9] Jakovljević S, Lisjak D, Alar Ž, Penava F. The influence of humidity on mechanical properties of bamboo for bicycles. Construction and Building Materials. 2017;**150**:35-48

[10] Gu X, Deng X, Liu Y, Zeng Q, Wu X, Ni Y, Liu X, Wu T, Fang P, Wang B, Wu Q. Review on comprehensive utilization of bamboo residues. Transactions of the Chinese Society of Agricultural Engineering. 2016;**32**(1):236-242

[11] Nayak L, Mishra SP. Prospect of bamboo as a renewable textile fiber, historical overview, labeling, controversies and regulation. Fashion and Textiles. 2016;**3**(1):2

[12] Chaowana P, Barbu MC. Bamboo: Potential material for biocomposites. In: Lignocellulosic Fibre and Biomass-Based Composite Materials; 2017. pp. 259-289

[13] Penellum M, Sharma B, Shah DU, Foster RM, Ramage MH. Relationship of structure and stiffness in laminated bamboo composites. Construction and Building Materials. 2018;**165**:241-246

[14] Nurul Fazita MR, Jayaraman K, Bhattacharyya D, Mohamad Haafiz MK, Saurabh CK, Hussin MH, Abdul Khalil HPS. Green composites made of bamboo fabric and poly (lactic) acid for packaging applications—A review. Materials. 2016;**9**(6):435

[15] Choudhury DSJK, Sharma GD. Value addition to bamboo shoots: A review. Journal of Food Science and Technology. 2011

[16] Gielis J. Future Possibilities for Bamboo in European Agriculture. Sint-Lenaartsesteenweg Rijkevorsel: Oprins Plant; 2002 91 (B-2310)

[17] Van der Lugt P, Van den Dobbelsteen AAJF, Janssen JJA. An environmental, economic and practical assessment of bamboo as a building material for supporting structures. Construction and Building Materials. 2006;**20**(9):648-656

[18] Azmy HM, Norini H, Wan Razali WM. Management guidelines and economics of natural bamboo stands. In: FRIM Technical Information Handbook. Kepong, Malaysia: Forest Research Institute Malaysia (FRIM); 1997;**15**:40

[19] Azmy HM, Abd Razak O. Field identification of twelve commercial malaysian bamboos. In: FRIM Technical Information. Vol. 25. Kepong, Malaysia: Forest Research Institute Malaysia; 1991. p. 12

[20] Lobovikov M, Ball L, Guardia M, Russo L. World Bamboo Resources: A Thematic Study Prepared in the Framework of the Global Forest Resources Assessment 2005 (No. 18). Roma, Italy: Food & Agriculture Organization; 2007

[21] Abdul Khalil HPS, Alwani MS, Islam MN, Suhaily SS, Dungani R, H'ng YM, Jawaid M. The use of bamboo fibres as reinforcements in composites. In: Biofiber Reinforcements in Composite Materials; 2015. pp. 488-524

[22] Yu HQ, Jiang ZH, Hse CY, Shupe TF. Selected physical and mechanical properties of moso bamboo (*Phyllostachys pubescens*). Journal of Tropical Forest Science. 2008;**20**:258-263

[23] Jiang ZH. World Bamboo and Rattan. Shenyang: Liaoning Scientific and Technological Publishing House; 2002

[24] Meng FD, Yu YL, Zhang YM, Yu WJ, Gao JM. Surface chemical composition analysis of heat-treated bamboo. Applied Surface Science. 2016;**371**:383-390

[25] Begum PR, Reddy AR, Hussain P, Reddy BS, Babu SS. Design and Analysis of Short Bamboo Fiber Reinforced Composite Materials. 2015

[26] Lee AWC, Xuesong B, Perry NP. Selected physical and mechanical properties of giant timber bamboo grown in South Carolina. Forest Products Journal. 1994;**44**(9):40-46

[27] Xian XJSDG, Ye YW. Bamboo Material and Bamboo Fiber Composite Materials. Beijing: Science Press; 1995

[28] Li XB. Physical, chemical, and mechanical properties of bamboo and its utilization potential for fiberboard manufacturing [MSc thesis]. Baton Rouge: Louisiana State University; 2004

[29] Xie X, Zhou Z, Jiang M, Xu X, Wang Z, Hui D. Cellulosic fibers from rice straw and bamboo used as reinforcement of cement-based composites for remarkably improving mechanical properties. Composites Part B: Engineering. 2015;**78**:153-161

[30] Rosamah E, Khalil AH, Yap SW, Saurabh CK, Tahir PM, Dungani R, Owolabi AF. The role of bamboo nanoparticles in kenaf fiber reinforced unsaturated polyester composites. Journal of Renewable Materials. 2017;**6**:75-86

[31] Tamizi MM. Fundamental and characteristic study of cultivated Malaysian bamboo—Selective genus *Gigantochloa* [Ph. D Thesis]. Universiti Sains Malaysia; 2011

[32] Liese W. Anatomy and properties of bamboo. Recent research on bamboos. In: Proceedings of the International Bamboo Workshop; 6-14 October 1985; Hangzhou, China

[33] Huang P, Chang WS, Ansell MP, Chew YJ, Shea A. Density distribution profile for internodes and nodes of *Phyllostachys edulis* (Moso bamboo) by computer tomography scanning. Construction and Building Materials. 2015;**93**:197-204

[34] Engler B, Dietenberger R, Schönherr S, Zhong Z, Becker G. Suitability of bamboo as an energy resource: Analysis of bamboo calorific value in dependency on the culms age. In: Proceedings Venice 2010, Third International Symposium on Energy from Biomass and WasteVenice; 8-11 November 2010; Italy

[35] Kamthai S. Preliminary study of anthraquinone in sweet bamboo (*dendrocalamus asper* backer) alkaline sulfite pulping. Chiang Mai Journal of Science. 2007;**34**(2):235-247

[36] Chen H, Menghe M, Ding X. Chemical treatments of bamboo to modify its moisture absorption and adhesion to vinyl ester resin in humid environment. Journal of Composite Materials. 2011

[37] Janssen JJA. The relationship between mechanical properties and the biological and chemical composition of bamboo. In: Higuchi T, editors. Bamboo Production and Utilization. Proceedings of XVIII IUFRO World Congress; 1981; Kyoto. Kyoto, Japan: Kyoto University; pp. 27-32

[38] Espiloy ZB. Some Properties and Uses of Bamboo. FORPRIDECOM Tech. Note No. 109. Philippines: Laguna: Forest Products Research and Industries Development Commission; 1987. p. 25

[39] Hamdan H, Anwar UMK, Zaidon A, Tamizi MM. Mechanical properties and failure behavior of bamboo (*Gigantochloa scortechinii*). Journal of Tropical Forest Science. 2009;**21**:4

[40] Panshin AJ, De Zeeuw C. Text Book of Wood Technology. Volume I. New York: McGraw-Hill Book Co; 1970. p. 705

[41] Amada S, Untao S. Fracture properties of bamboo. Composites: Part B. 2001;**32**:451-459

[42] Xu M, Cui Z, Chen Z, Xiang J. Experimental study on compressive and tensile properties of a bamboo scrimber at elevated temperatures. Construction and Building Materials. 2017;**151**:732-741

[43] Latif AM, Ashaari A, Jamaludin K, Zin JM. Effects of anatomical characteristics on the physical and mechanical properties of Bambusa Bluemeana. Journal of Tropical Forest Science. 1993;**61**:59-170

[44] Sattar MA, Kabir MF, Bhattacharjee DK. Effect of age and height position of muli (*Melocanna baccifera*) and borak (*Bambusa balsooa*) bamboo on their physical and mechanical properties. In: Proceedings: 4th International Bamboo Workshop. Bamboo in Asia and the Pacific; Chiangmai, Thailand. 1991

[45] Sattar MA, Kabir MF, Bhattacharjee DK. Physical and mechanical properties of *Bambusa arundinacea, Bambusa longispiculata, Bambusa vulgaris* and *Dendrocalamus giganteus*. Bangladesh Journal of Forest Science. 1994;**23**:20-25

[46] Espiloy ZB, Ella AB, Floresca AR. Physico-mechanical properties and anatomical structure relationships of two erect bamboo species. The Philippines Lumberman. 1986;**32**:25-27

[47] Latif AM, Wan Tarmeze WA, Fauzidah A. Anatomical features and mechanical properties of three Malaysian bamboos. Journal of Tropical Forest Science. 1990;**2**:227-234

[48] Janssen JJA. Mechanical Properties of Bamboo. Vol. 37. Kluwer Academic Publishers, Forestry Sciences; 1991

[49] Lakkad SC, Patel JM. Mechanical properties of bamboo, a nature composite. Fibre Science and Technology. 1981;**14**:319-322

[50] Defoirdt N, Biswas S, De Vriese L, Tran LQN, Acker JV, Ahsan Q, Gorbatikh L,Vuure AV ,Verpoest I. Assessment of the tensile properties of coir, bamboo and jute fibre. Composites: Part A. 2010;**41**:588-595

[51] Tan T, Rahbar N, Allamch SM, Kwofie S, Dissmore D, Ghavami K, Soboyejo WO. Mechanical properties of functionally graded hierarchical bamboo structures. Acta Biomaterialia. 2011

[52] Rao KMM, Rao KM. Extraction and tensile properties of natural fibers: Vakka, date and bamboo. Composite Structures. 2007;**77**:288-295

[53] Espiloy ZB, Sasondoncillo RS. Some characteristics and properties of giant bamboo (Gigantochioa aspera Kurz). Forpride Digest. 1978;**7**:34-36

[54] Biswas D, Bose S, Hossain MM. Physical and mechanical properties of urea formaldehyde-bonded particleboard made from bamboo waste. International Journal of Adhesion and Adhesives. 2011;**31**:84-87

[55] Li SH, Liu Q, de Wijn JR, Zhout BL, de Groat K. In vitro calcium phosphate formation on a natural composite material, bamboo. Biomaterials. 1997;**18**(5)

[56] Ghavami K, Rodrigues CS, Paciornik S. Bamboo: Functionally graded composite material. Asian Journal of Civil Engineering (Building and Housing). 2003;**4**:1

[57] Lee SM, Cho D, Park WH, Lee SG, Han SO, Drzal LT. Novel silk/poly(butylene succinate) biocomposites: The effect of short fibre content on their mechanical and thermal properties. Composites Science and Technology. 2005;**65**:647-657

[58] Fushimi C, Katayama S, Tsutsumi A. Elucidation of interaction among cellulose, lignin and xylan during tar and gas evolution in steam gasification. Journal of Analytical and Applied Pyrolysis. 2009;**86**:82-89

[59] Teng H, Wei YC. Thermogravimetric studies on the kinetics of rice hull pyrolysis and the influence of water treatment. Industrial and Engineering Chemistry Research. 1998;**37**:3806-3811

[60] Xie H, King A, Kilpelainen I, Granstrom M, Argyropoulos DS. Thorough chemical modification of wood-based lignocellulosic materials in ionic liquids. Biomacromolecules. 2007;**8**:3740-3748

[61] Mui ELK, Cheung WH, Lee VKC, McKay G. Kinetic study on bamboo pyrolysis. Industrial and Engineering Chemistry Research. 2008;**47**:5710-5722

[62] Jeguirim M, Dorge S, Trouve G. Thermogravimetric analysis and emission characteristics of two energy crops in air atmosphere: *Arundo donax* and *Miscanthus giganthus*. Bioresource Technology. 2010;**101**:788-793

[63] Gomez CJ, Meszaros E, Jakab E, Velo E, Puigjaner L. Thermogravimetry/mass spectrometry study of woody residues and an herbaceous biomass crop using PCA techniques. Journal of Analytical and Applied Pyrolysis. 2007;**80**:416-426

[64] Nabi SD, Jog JP. Natural fiber polymer composites: A review. Advances in Polymer Technology. 1999;**18**:351-363

[65] Munir S, Daood SS, Nimmo W, Cunliffe AM, Gibbs BM. Thermal analysis and devolatilization kinetics of cotton stalk, sugar cane bagasse and shea meal under nitrogen and air atmospheres. Bioresource Technology. 2009;**100**:1413-1418

[66] Barneto AG, Carmona JA, Alfonso JEM, Serrano RS. Simulation of the thermogravimetry analysis of three non-wood pulps. Bioresource Technology. 2002;**101**:3220-3229

[67] Gronli MG, Vahegyi G, Di Blasi C. Thermogravimetric analysis and devolatilization kinetics of wood. Industrial and Engineering Chemistry Research. 2002;**41**:4201-4208

[68] Jauhiainen J, Conesa JA, Font R, Martan-Gullan I. Kinetics of the pyrolysis and combustion of olive oil solid waste. Journal of Analytical and Applied Pyrolysis. 2004;**72**:9-15

[69] Yang H, Yan R, Chen H, Lee DH, Zheng C. Characteristics of hemicellulose, cellulose and lignin pyrolysis. Fuel. 2007;**86**:1781-1788

[70] Font R, Marcilla A, Verda E, Devesa J. Thermogravimetric kinetic study of the pyrolysis of almond shells and almond shells impregnated with CoCl. Journal of Analytical and Applied Pyrolysis. 1991;**21**:249-264

[71] Liu NA, Fan W, Dobashi R, Huang L. Kinetic modeling of thermal decomposition of natural cellulosic materials in air atmosphere. Journal of Analytical and Applied Pyrolysis. 2002;**63**:303-325

[72] Das M, Chakraborty D. Thermogravimetric analysis and weathering study by water immersion of alkali treated bamboo strips. BioResources. 2008;**3**:1051-1062

[73] Rajulu AV, Reddy GR, Chary KN, Rao GB, Devi LG. Thermogravimmetric analysis of Dendrocalamus strictus bamboo fibres. Journal of Bamboo and Rattan. 2002;**1**:247-250

[74] Khalil HA, Hossain MS, Rosamah E, Norulaini NN, Leh CP, Asniza M, Davoudpour Y, Zaidul ISM. High-pressure enzymatic hydrolysis to reveal physicochemical and thermal properties of bamboo fiber using a supercritical water fermenter. BioResources. 2014;**9**(4):7710-7720

[75] Liew FK, Hamdan S, Rahman M, Rusop M, Lai JCH, Hossen M. Synthesis and characterization of cellulose from green bamboo by chemical treatment with mechanical process. Journal of Chemistry. 2015

[76] Pai Peng FP, Bian J, Feng X, Sun R. Studies on the starch and hemicelluloses fractionated by graded ethanol precipitation from bamboo phyllostachys bambusoides f. shouzhu Yi. Journal of Agricultural and Food Chemistry. 2011;**59**:2680-2688

[77] Nowakowski DJ, Jones JM. Uncatalysed and potassiumcatalysed pyrolysis of the cell-wall constituents of biomass and their model compounds. Journal of Analytical and Applied Pyrolysis. 2008;**83**:12-25

[78] Krzesinska M, Zachariasz J, Muszynski J, Czajkowska S. The thermal decomposition studies of solid iron bamboo (*Dendrocalamus strictus*)—Potential precursor for eco-materials. Bioresource Technology. 2008;**99**:5110-5114

Chapter 10

A Review of Bambusicolous Ascomycetes

Dong-Qin Dai, Li-Zhou Tang and Hai-Bo Wang

Additional information is available at the end of the chapter

http://dx.doi.org/10.5772/intechopen.76463

Abstract

Bamboo with more than 1500 species is a giant grass and was distributed worldwide. Their culms and leaves are inhabited by abundant microfungi. A documentary investigation points out that more than 1300 fungi including 150 basidiomycetes and 800 ascomycetous species with 240 hyphomycetous taxa and 110 coelomycetous taxa are associated with bamboo. Ascomycetes are the largest group with totally 1150 species. Families *Xylariaceae* and *Hypocreaceae*, which are most represented, have 74 species and 63 species in 18 and 14 genera, respectively, known from bamboo. The genus *Phyllachora* with a maximum number of species (22) occurs on bamboo, followed by *Nectria* (21) and *Hypoxylon* (20). The most represented host genera *Bambusa*, *Phyllostachys*, and *Sasa* are associated by 268, 186, and 105 fungal species, respectively. The brief review of major morphology and phylogeny of bambusicolous ascomycetes is provided, as well as research prospects.

Keywords: bamboo fungi, pathogens, morphology, phylogeny

1. Introduction

Bamboo, as the largest member of the grass family *Poaceae*, plays an important role in local economies throughout the world. They are used in furniture and construction, or as food for human and animals like panda. Bamboo is even used in Chinese traditional medicine for treating infections and healing of wounds [1]. Bamboo species is distributed in diverse climates, from cold mountain areas to hot tropical regions. They have very little natural toxicity and, therefore are easily prone to fungi and insect attacks [2]. It is reported that more than 1100 species of fungi have so far been described or recorded from bamboo host worldwide [3]. These include ca. 630 ascomycetes, 150 basidiomycetes, and 330 asexual morph taxa (100 coelomycetous and 230 hyphomycetous species) [3]. Bamboo fungi are important to agricultural and economic development, such as *Phallus indusiatus* Vent., which is the delicious edible mushroom and usually called as "bamboo pith." It has been cultivated worldwide and brings a high economic income. Some of bamboo fungi, however, are pathogens, most of which are ascomycetes. In this chapter, a review of bamboo ascomycetes is provided herein.

2. Importance of bamboo ascomycetes

Cultivation of economically important bamboos is often threatened by fungal infection and diseases which eventually result from serious damages on bamboo cultivation [4]. A number of foliage diseases (e.g., leaf spots and leaf blight) of bamboos have been recorded. However, fungi cause comparatively less damage to bamboo than culm diseases [5]. For example, during 1988–1990, 5300 hectares of bamboos were affected by *Balansia take* (Miyake) Hara which occurred on bamboo culms in Fujian, China. More than 200,000 bamboos were cut down and burned to prevent the pathogen from spreading. However, this caused serious economic losses. In 1960, Hino and Katumoto explained that the relative importance of different diseases affecting bamboos is difficult to assess because of the general lack of information accompanying the disease records [4]. Thus, pathologists have now started paying attention on the research of bamboo pathogens. A book entitled *Diseases of bamboos in Asia: an illustrated manual* listed 122 fungal diseases in Asia region, with 100 records in India and 15 in Thailand [5]. In China, 183 fungal pathogens associated with bamboo were recorded [6, 7]. Numbers of ascomycetous fungi recorded as bamboo pathogens were quite high. However, in the absence of molecular data, most of the fungal names recognized remained artificial. A sizable number of taxa were not even identified up to species level.

Shiraia bambusicola Henn. is one of the famous bamboo pathogens because of its medicinal value. The fungus produces large, pinkish ascostromata on living bamboo branches (**Figure 1**). Their fruiting bodies, as a traditional Chinese medicine, are used for curing rheumatoid arthritis [8], as well as extracting the metabolite, hypocrellin [9], which has promising applications

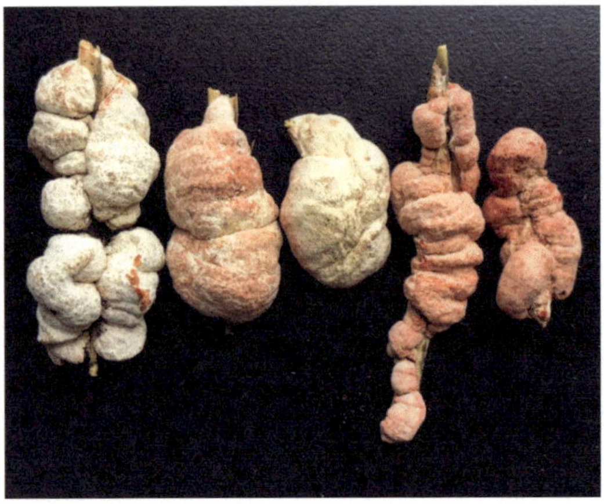

Figure 1. Ascostromata of *Shiraia bambusicola* collected from China.

in photodynamic therapy (PDT) [10] and in anticancer treatments [11]. Another well-known Chinese medicinal ascomycete is *Hypocrella bambusae* (Berk. & Broome) Sacc. This fungus produces similar ascostromata with *S. bambusicola* on branches of *Sinarundinaria* spp. Their ascostromata are also used to extract hypocrellin A and hypocrellin B. It is reported that hypocrellin B can be used to evaluate antiviral activity against the human immunodeficiency virus (HIV-1) [12]. In China, a costly medicinal unguent named Bamboo Parasitic Fungus Ointment is made of hypocrellin B. *H. bambusae* contains higher hypocrellin than that of *S. bambusicola*. Index Fungorum [13] recorded that the current name of this fungus is *Pseudonectria bambusae* (Berk. & Broome) Höhn. Without the full morphological study and molecular data in GenBank, its taxonomic placement still remains confused. Therefore, the study of taxonomy and phylogeny of bamboo fungi is urgently needed to be carried out to clarify these undetermined and confused taxa.

3. Bamboo ascomycetes: history

The term "fungorum bambusicolorum" (bambusicolous fungi), was first used by Iwao Hino [14], though the author did not give a definition. "Bambusicolous" means "living on bamboo" [3]. Kevin D. Hyde and colleagues in 2002 defined bambusicolous, which embodies fungi growing on any bamboo substrates, including leaves, culms, branches, sheathes, flowers, rhizomes, and roots [3]. Subsequently, the phrase "Bambusicolous fungi" has been widely used by mycologists worldwide [15, 16].

Lembosia Léveillé is the first mycologist, who mentioned the presence of a fungus on bamboo. In 1845, he described *Roumegueria goudotii* (Lév.) Sacc. ex Clem. & Shear (Basionym: *Dothidea goudotii* Lev.) occurring on leaves of *Chusquea* sp. and *Sphaeria bambusae* Lev. collected from the culms of *Bambusa bambos* (L.) Voss, in Tolima, Columbia [17]. Later, this author introduced two new species *Asterina microscopica* Lev. and *S. hypoxantha* Lev. from the leaves of *Chusquea* sp. and culms of *B. bambos*, respectively, in 1846 [18].

In 1854, Miles J. Berkeley recorded *Hypoxylon fuscopurpureum* (Schwein.) Berk. from *Phyllostachys* sp. and *Sasa* sp. [19]. The next year, Jean P.F.C. Montagne introduced a new species *S. fusariispora* Mont. on leaves of *Bambusa* sp. [20]. During 1871–1880, eight new ascomycetes were described from bamboo host, and between 1881 and 1920, the numbers got increased up to more than 100 species (**Figure 2**). However, the number of newly described species declined before and after the Second World War [3] (**Figure 2**). Nevertheless, during 1951–1990, publications of new species on bamboo got increased dramatically (**Figure 2**). Iwao Hino and Ken Katumoto made an earlier significant contribution on bambusicolous fungi during 1960–1970s, by recording 104 new species of ascomycetes [4, 21–29]. In 1961, Iwao Hino wrote *Icones fungorum bambusicolorum japonicorum*, and recorded 460 ascomycetous species worldwide, of which 175 species were from Japan [30]. In the following years, more records and checklists of fungi on bamboo were carried out by mycologists. Petrini Orlando and colleagues in 1989 mentioned 63 records from France [31]. Ove E. Eriksson and Yue Jinzhu published *Bambusicolous pyrenomycetes, an annotated checklist* and listed 587 ascomycetous taxa in 1998 [32]. A checklist for 104 species from China is provided in 1999 by [33].

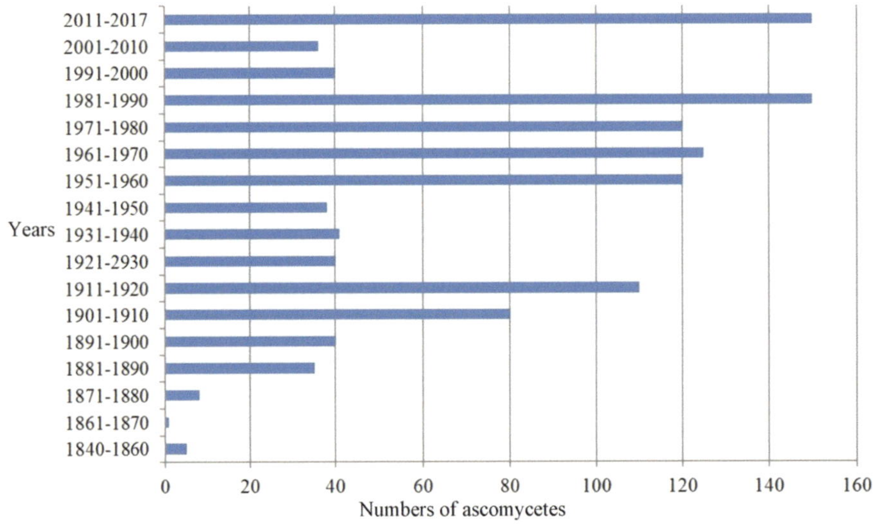

Figure 2. Number of fungi described from bamboo until 2017.

Recent contributions to the knowledge of bamboo-associated ascomycetes were the following. An article in 2000 reviewed 189 species of fungi, of which were 80 ascomycetes, 40 coelomycetous, and hyphomycetous fungi on bamboo substrates [34]. Kevin D. Hyde and colleagues reviewed bamboo fungi in detail in 2002 [3]. A total of 80 fungi were recorded on submerged bamboo in 2003, see [35].

During 2003–2008, the Japanese mycologists carried out a series of studies on bambusicolous fungi in Japan and introduced around 25 new taxa based on the morphological characters [15, 36–42]. In 2007, authors analyzed molecular diversity of bamboo-associated fungi from Japan, based on 257 endophytes strains, isolated from bamboo tissues; however, most isolates were not identified to species level [43]. Until 2009, the significant phylogenetic analysis was first used to classify the new taxa by [44].

Currently, during 2011–2017, approximate 145 new species and new records belonging to 65 genera, 37 families have so far been described or reported by mycologists [16, 45–52] based on taxonomic and phylogenetic analyses. In the short period of 6 years, six new families *viz. Anteagloniaceae* K.D. Hyde & Mapook [53], *Bambusicolaceae* D.Q. Dai & K.D. Hyde [54], *Occultabambusaceae* D.Q. Dai & K.D. Hyde [3], *Parabambusicolaceae* Kaz. Tanaka & K. Hiray [52], *Pseudoastrosphaeriellaceae* Phookamsak & K.D. Hyde [55], *Roussoellaceae* J.K. Liu et al. [56]. and 21 new genera viz. *Amphibambusa* D.Q. Dai & K.D. Hyde, *Bambusaria* Jaklitsch et al., *Bambusicola* D.Q. Dai & K.D. Hyde, *Bambusistroma* D.Q. Dai & K.D. Hyde, *Botryobambusa* Phook., J.K. Liu & K.D. Hyde, *Brunneoclavispora* Phook. & K.D. Hyde, *Embryonispora* G.Z. Zhao, *Flammeascoma* Phook. & K.D. Hyde, *Gregarithecium* Kaz. Tanaka & K. Hiray, *Kalmusibambusa* Phookamsak, Tennakoon, *Multilocularia* Phook., Ariyaw. & K.D. Hyde, *Neoanthostomella* D.Q. Dai & K.D. Hyde, *Neogaeumannomyces* D.Q. Dai & K.D. Hyde, *Neoramichloridium* Phookamsak et al., *Neoophiosphaerella* Kaz. Tanaka & K. Hiray., *Neoroussoella* J.K. Liu, Phook.

& K.D. Hyde, *Occultabambusa* D.Q. Dai & K.D. Hyde, *Parabambusicola* Kaz. Tanaka & K. Hiray, *Pseudoastrosphaeriella* Phookamsak et al., *Pustulomyces* D.Q. Dai & K.D. Hyde, *Seriascoma* Phook., D.Q. Dai & K.D. Hyde, associated with bamboo substrates were introduced by the natural classification [16, 45, 46, 52, 53, 55, 57–62].

4. Major morphological characters of bamboo ascomycetes

The morphological characters are important in fungal identifications. Most fungi produce their fruiting bodies on hyphae with two phases of reproductions (sexual and asexual reproduction). Thus, when an ascomycetous fungus bears its fruiting body by sexual reproduction, it usually produces an ascoma (**Figure 3**), and on the other hand, it produces either a conidioma or hyphomycetous fruiting structures. The major morphological characters of sexual morph are the type of ascomata, hamathecium, asci, and ascospores, and those of asexual morph are conidiomata, conidiophores, and conidiogenous cells [63, 64].

4.1. Morphological characters of sexual morph

4.1.1. Ascomata

The ascomata of bambusicolous fungi have various types, no matter what shapes or colors. They can form on bamboo leaves, culms, or even sheathes (**Figure 4**), with the positions being immersed, erumpent, or superficial (**Figure 5**). Some ascomata are stromatic, with uniloc-ule to multilocules. Ascomata produce asci and ascospores within, when mature. An ostiole, where ascospores release, is commonly present at the top of an ascoma. Peridium is the wall of an ascoma and usually is composed of several layers of angular cells. The hamathecium is the sterile tissue, formed by hyaline filaments which are called paraphyses.

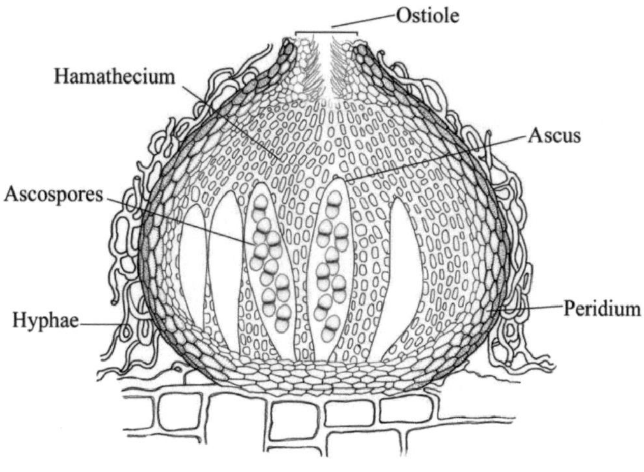

Figure 3. Line drawing of a fruiting body (ascoma) of sexual morph of ascomycete.

Figure 4. Ascomata on different bamboo substrates. (a), (b) on bamboo leaves. (c) on bamboo sheath.(d) on bamboo culm.

4.1.2. Asci

The morphology of asci is particularly of two major types, unitunicate and bitunicate (**Figure 6**). Asci are produced within the ascomata and act as platforms from which the spores are launched. However, some asci release the ascospores passively by dissolving the ascal wall at maturity, rather than active spore-shooting mechanism. Asci usually bear eight ascospores, though sometimes with four or six, and occasionally producing multiple numbers of ascospores.

4.1.3. Ascospores.

The ascospores of bambusicolous fungi exhibit a variety of shapes (**Figure 7**). The common shapes are ellipsoidal, fusiform, filiform, and so on. The color of ascospores can be hyaline, pale brown, brown to dark brown. The ascospores usually are single-celled or have one to two septa and sometimes multiseptate, like muriform spores. Their surfaces are smooth, striate to verrucose, or covered by a sheath.

4.2. Morphological characters of asexual morph.

The asexual morph of ascomycetous bambusicolous fungi is coelomycetous or hyphomycete-ous [45–48], as in the asexual morph of *Bambusicola* and *Apiospora* Sacc. (≡*Arthrinium* Kunze)

Figure 5. Various types of ascomata on bamboo. a–g, l: superficial ascomata; h–j: erupted ascomata; k: immersed ascomata; a–c, e, h: perithecia; d: stromata; f, g: flatten ascostromata; i: multiloculate stroma; j: conical stromatic ascomata; k: stromata; l: *xylarioid* stromata.

[16, 45]. The major morphologic characters are conidiomata, conidiophores, and conidiogenous cells. In the coelomycetous fungi, the conidia develop in a growing cavity, called conidiomata which can be acervuli, pycnidia, or sporodochium-like structures [64]. For hyphomycetous fungi, a conidium develops directly on the conidiophores which may be mononematous or synnematous and bear a single or more conidiogenous cells which usually are holoblastic, enteroblastic, phialidic, annelidic, or tretic. Conidium ontogeny has long been used as an

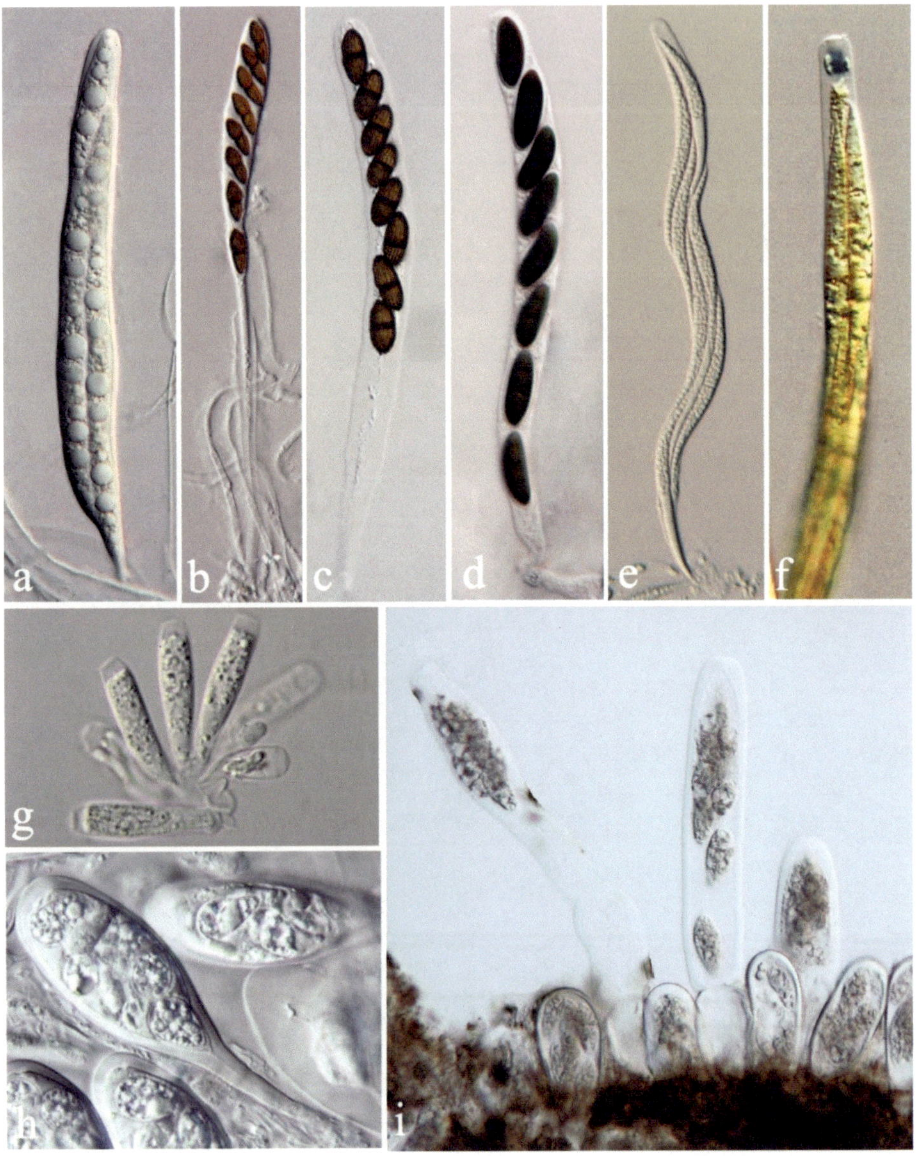

Figure 6. Asci types. a, b, c, h: bitunicate asci (note in b, jack-in-a-box spore release mechanism). d, e: unitunicate asci; f: unitunicate ascus with a special elastic ring at the tip; g: unitunicate asci with a flat apex; h: bitunicate ascus with a long narrow basal portion; i: fissitunicate asci.

important taxonomic feature in the diagnosis of conidial fungi [65]. Mostly, the conidiophores of coelomycetous fungi are reduced to conidiogenous cells, when they grow on the host substrate. However, some of them remain hyphomycetous (producing free conidiophores) on culture, as in *Arthrinium* [51].

Figure 7. Ascospores types. a, e: ascospores surrounded by a sheath; b: striate ascospore; c, k: verrucose ascospores; d: dark brown, 6-septate ascospore; f, i, j: fusiform ascospores; g: muriform ascospore; h: smooth-walled ascospore; l, m: filiform ascospores.

5. Bamboo ascomycetes taxonomic distribution

The ascomycetes on bamboo are very diverse, together with their asexual morphs. In [30], the author recorded 460 species belonging to 218 genera in 43 families from all over the world. A reference [3] reported 630 species distributed in 121 families and 436 genera within ascomycota. According to our investigation on the basis of the current references, 800 ascomycetous species (**Figure 8**) belonging to 465 genera within 128 families were documented.

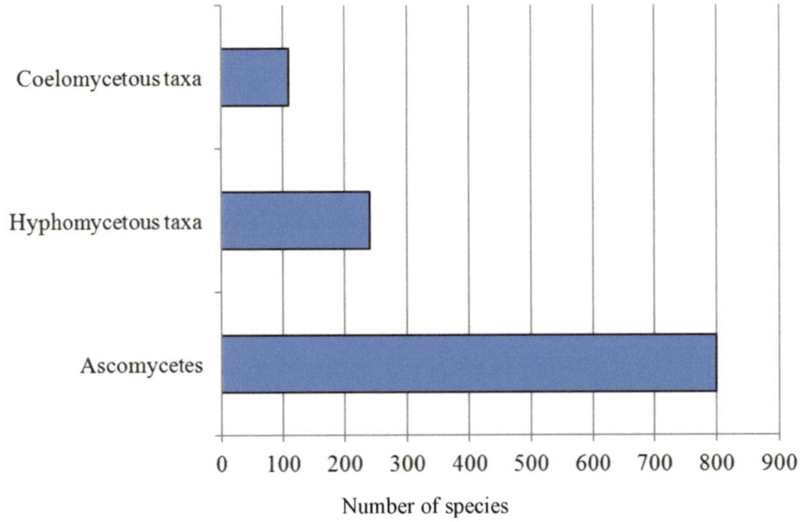

Figure 8. Numbers of ascomycetous, coelomycetous, and hyphomyceteous species on bamboo.

Families of *Xylariaceae* and *Hypocreaceae,* which are most represented, have 74 species and 63 species in 18 and 14 genera, respectively, known from bamboo (**Figures 9** and **10**). This was followed by the families, *Phyllachoraceae* and *Lasiosphaeraceae,* with 35 species in 9 genera and 33 species in 10 genera, respectively. *Roussoellaceae* with 32 species, *Clavicipitaceae* with 28

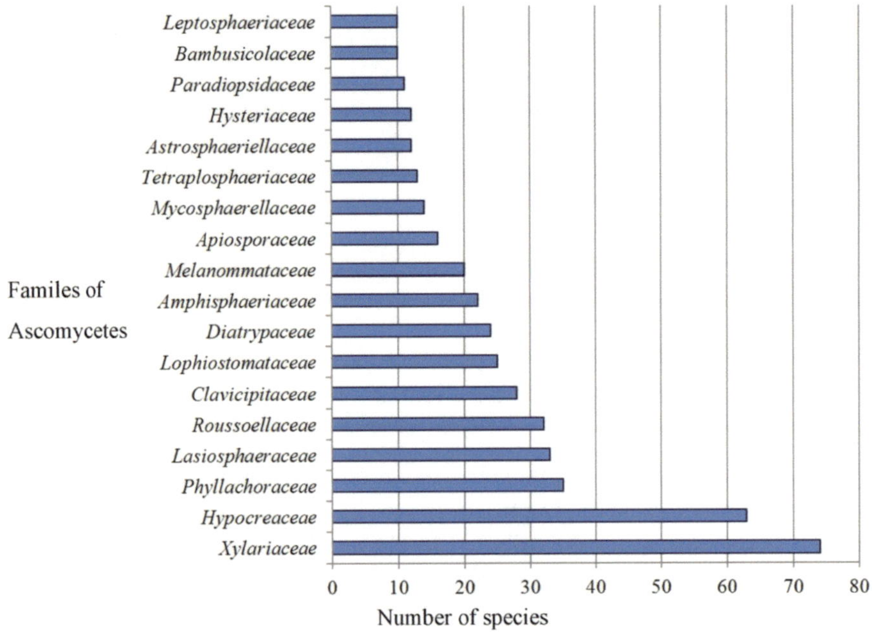

Figure 9. Ascomycete families with more than 10 recorded species on bamboo until 2017.

http://dx.doi.org/10.5772/intechopen.76463

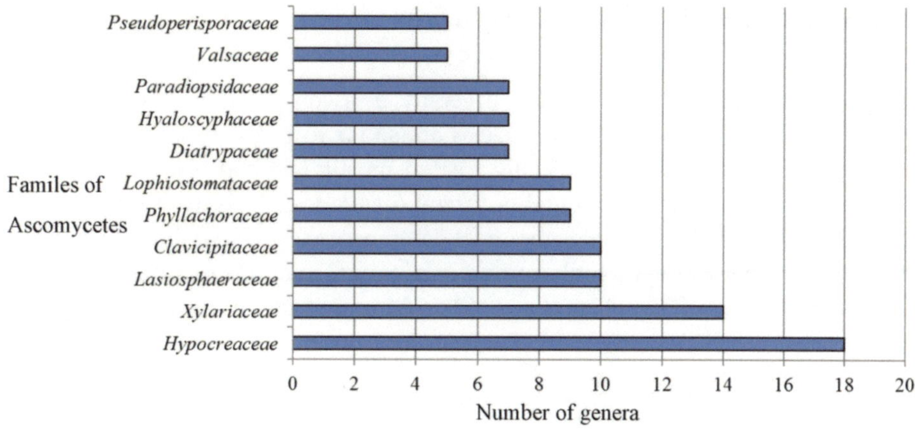

Figure 10. Ascomycete families with more than five genera recorded on bamboo until 2017.

species, *Lophiostomataceae* with 25 species, *Diatrypaceae* with 24 species, *Amphisphaeriaceae* with 22 species, and *Melanommataceae* with 20 species are the next common families on bamboo (**Figure 9**).

The genus *Phyllachora* with a maximum number of species (22) occurs on bamboo, followed by *Nectria* (21) and *Hypoxylon* (20) (**Figure 11**). *Phyllachora* species are the commonly known fungi on the grass families [3]. *Rosellinia* has hundreds of epithets listed, with 18 species occurring on bamboo. *Roussoella* is a well-known genus on bamboo and palm hosts, with 17 species known from the bamboo. Genera such as *Didymosphaeria* with 15 species, *Anthostomella* with 14, *Arthrinium* with 16, *Hypocrea* with 11, and *Astrosphaeriella* and *Chaetosphaeria* with 10 are also seen on bamboo. *Bambusicola* is a new genus introduced by [45] and contains 10 species as well found on bamboo culms during 2012–2017.

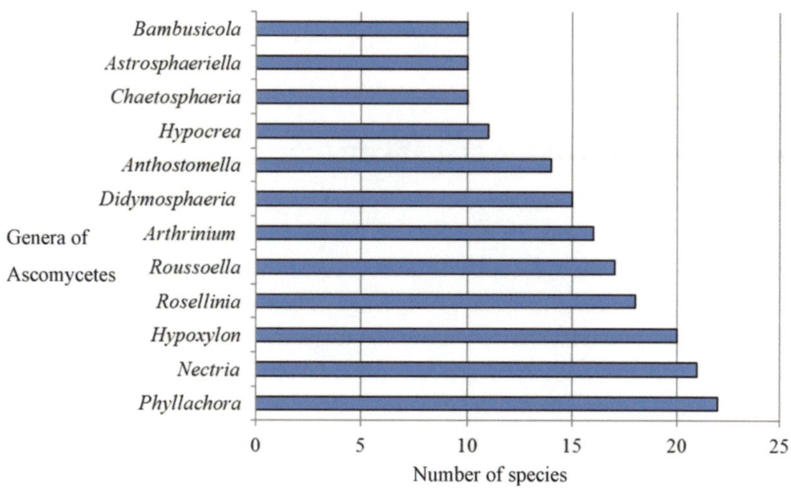

Figure 11. Ascomycetous genera with more than 10 recorded species on bamboo until 2017.

6. Bamboo ascomycetes: host distribution

Figure 12 showcases a picture of higher diversity of ascomycetes on the substrate genus *Bambusa*. More than 260 ascomycetous species are so far recorded from host *Bambusa*. This may be due to that the authors artificially treated mostly bamboo host names as *Bambusa* spp. The next bamboo genus with a higher fungal diversity is *Phyllostachys* with around 180 fungal species. *Phyllostachys* is one of the most common genera of bamboo. This may be the possible reason why it yielded a higher number of fungal taxa. There are 105 ascomycetous species occurring on *Sasa*, 73 on *Arundinaria*, 60 on *Chusquea*, and 56 on *Pleioblastus*. Such high species diversity at the subfamily level (*Bambusoideae*) undoubtedly has a significant impact on species numbers [3].

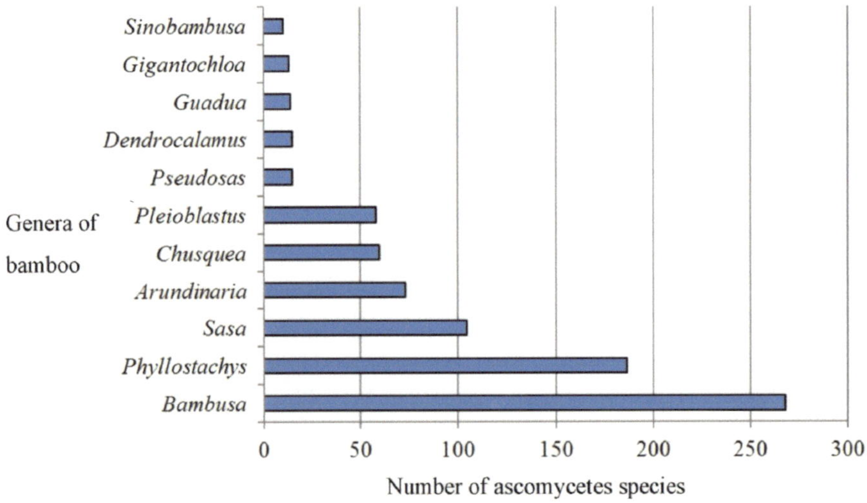

Figure 12. Bamboo genera with more than 10 recorded ascomycetous species until 2017.

7. Bamboo ascomycetes: tissue specificity

It is reported that most fungi grow on bamboo culms (514 fungal species) and leaves and only few recorded species from shoots, roots, or inflorescences [3]. An update of bamboo tissue types with their fungal species number is shown in **Figure 13** and according to which 665 fungal species have been collected from bamboo culms and 216 species were recorded from leaves, followed by sheaths (19 species) and branches (14 species). It is unknown whether fungi are tissue specific or are simply recurrent on certain bamboo tissues [3]. Most bamboo pathogens affect leaves, although a large number of ascomycetes (e.g., *Roussoella* spp. and *Astrosphaeriella* spp.) have been reported on decaying culms or branches.

http://dx.doi.org/10.5772/intechopen.76463

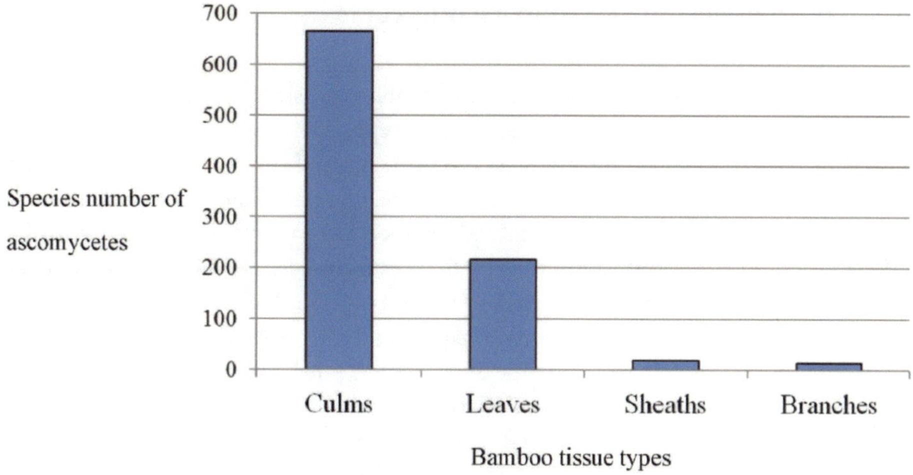

Figure 13. Bamboo ascomycetes: distribution by tissue types until 2017.

8. Phylogeny of bamboo ascomycetes

Phylogenetic analysis has been widely used to study the evolution and relationships among fungal individuals or groups. It is the modern method to clarify fungal species or fungal higher level placements (viz. genus, family, order, etc.). Initially, the studies of bamboo ascomycetes were carried out only based on the morphologic examinations. At the beginning of the twenty-first century, Japanese mycologists made the first contributed acknowledge on bamboo ascomycetes identifications and introduced a new fungal family to accommodate five new genera founded on bamboo [52]. To evaluate the validity of fungal taxa and clarify their phylogenetic relationships, the combined sequences of data with multi-genes (LSU (large subunit rDNA), SSU (small subunit rDNA), TEF (translation elongation factor 1-α gene region), and beta-tubulin) were firstly used to make the phylogenetic trees for bamboo-associated ascomycetes in [16]. Following, more new species, genera, and several new families were established by Thai and Chinese researchers based on morphology-phylogeny and the linking of sexual-asexual morphs [45, 55]. In their studies, ITS (internal transcribed spacers) and beta-tubulin genes were usually used to indentify fungal species like species in *Arthrinium* and *Myrothecium* [16, 51] and LSU, SSU, TEF, and RPB2 (RNA polymerase II second largest subunit) were used to clarify the relationships among genera, families, and orders [16, 52]. The evolution study of bamboo ascomycetes has not been carried out yet, probably due to most old species having no molecular data. More than 800 ascomycetous species with 240 hyphomycetous taxa and 110 coelomycetous taxa are so far recorded on bamboo; however, only less than 180 species have available sequences. Fresh specimens are required to recollect, and the existing species are waiting for reexamination or epitypification.

9. Research prospects of bamboo ascomycetes

Currently, 1300 fungi have been recorded on bamboo substrates. These are including 150 basidiomycetes [3], 800 ascomycetous species, 240 hyphomycetous taxa, and 110 coelomycetous taxa. It provides large opportunities on the utilization of fungi on bamboo. For example, species in *Hypoxylon* Bull., which form large stromata, can be easier to see on bamboo culms. In the recent years, a new secondary metabolite named lenormandins, which are further representatives of the well-known azaphilones [66], was found from *Hypoxylon* spp. More than 18 species in *Hypoxylon* have been occurred so far on bamboo (**Figure 11**). There may be some other interesting unknown azaphilones in *Hypoxylon* species, which may still wait for extraction. Numerous fungi from bamboo were documented, and some pathogens were isolated as well, such as *Arthrinium phaeospermum* (Corda), M.B. Ellis, and *Phyllachora graminis* (Pers.) Fuckel. Approximately 300 bamboo fungal diseases were so far reported [5–7, 67], which provides the essential step to understand ecosystem communication [3]. It is showed that some bamboo ascomycetes have a high value in medical treatment, such as *Engleromyces goetzi, H. bambusae,* and *S. bambusicola*. However, their effective compounds can be only obtained from stromata, which cannot be cultivated successfully at the moment [68]. The phylogenic relationship between *H. bambusae* and *S. bambusicola* needs to be carried out to understand the reason why *H. bambusae* produces more hypocrellin than the latter. Many interesting and subsequent studies on bamboo ascomycetes are still awaiting and needs to be conducted, and large numbers of new ascomycetous taxa on bamboo are also still waiting for collection and isolation.

10. Conclusions

Bamboo plays an important role in the forest ecosystem and is treated as an economic plant for human. Studying the bamboo fungi can provide the chances for controlling bamboo pathogens and promoting bamboo cultivation. Based on our study, more than 1300 bamboo ascomycetes have so far been described or recorded; however, most of them do not have detailed morphology or sequence data. Even some important ascomycetous species still need re-studying. The morphologic characters of bamboo ascomycetes are various in their ascomata, asci, and ascospores (**Figures 5–7**). They occur on different genera of bamboo host; however, most of the hosts have not been identified to species level. Bamboo ascomycetes are diverse in their taxonomic placements, with more than 120 families and 400 genera distributed according to the references. Phylogenic analyses of bamboo ascomycetes need more study and should focus on protein genes. It is recommended that more fresh specimens need to be collected in the future, and the existing species should be epitypified or designated as reference specimens. Efforts are required in naming the taxa to avoid confusion, such as *Hypocrella bambusae*. Host species names should be identified in the future for those willing to work on host specificity.

Acknowledgements

This work was supported by the Key Laboratory of Yunnan Province Universities of the Diversity and Ecological Adaptive Evolution for Animals and Plants on Yun-Gui Plateau, the

Scientific Research Foundation of Yunnan Provincial Department of Education (2017ZZX186), and the National Natural Science Foundation of China (Nos. NSFC 31760013, NSFC 31260087, NSFC 31460179, and NSFC 31460561). The authors would like to thank the Yunnan Province Universities of the Science and Technology Innovation Team for the exploitation and utilization of endophytes for finance support.

Author details

Dong-Qin Dai, Li-Zhou Tang* and Hai-Bo Wang

*Address all correspondence to: tanglizhou@163.com

Center for Yunnan Plateau Biological Resources Protection and Utilization, College of Biological Resource and Food Engineering, Qujing Normal University, Qujing, Yunnan, People's Republic of China

References

[1] Janssen JJ. Mechanical Properties of Bamboo. Netherlands: Kluwer Academic Publishers; 1991. pp. 30-113

[2] Opoku D, Ayarkwa J, Agyekum K. Alternative binders for increased sustainable construction in Ghana—A guide for building professionals. Materials Sciences & Applications. 2013;**4**(12):20-28

[3] Hyde KD, Zhou DQ, Dalisayl T. Bambusicolous fungi: A review. Fungal Diversity. 2002;**9**(1):1-14

[4] Hino I, Katumoto K. Illustrationes fungorum bambusicolorum VIII. Bulletin of the Faculty of Agriculture Yamaguti University. 1960;**11**:9-34

[5] Mohanan C. Diseases of Bamboos in Asia: An Illustrated Manual India. Leiden: Brill Academic Publishers; 2002. pp. 39-57

[6] Xu MQ, Dai YC, Fan SH, Jin LX, Lu Q, Tian GZ, Wang LF. Records of bamboo diseases and the taxonomy of their pathogens in China (I). Forest Research. 2006;**19**(6):692-699

[7] Xu MQ, Dai YC, Fan SH, Jin LX, Lu Q, Tian GZ, Wang LF. Records of bamboo diseases and the taxonomy of their pathogens in China (II). Forest Research. 2007;**20**(1):45-52

[8] Shen YX, Rong XG, Gao ZH. Studies on the chemical constituents of *Shiraia bambusicola*. China Journal of Chinese Materia Medica. 2002;**27**(9):674-677

[9] Zhen JD, Lown JW. Hypocrellins and their use in photosensitization. Photochemistry and Photobiology. 1990;**52**(6):609-616

[10] Dougherty TJ, Gomer CJ, Henderson BW, Jori G, Kessel D, Korbelik M, Moan J, Peng Q. Photodynamic therapy. Journal of the National Cancer Institute. 1998;**90**(12):889-905

[11] Liu YX, Hyde KD, Ariyawansa HA, Li WJ, Zhou DQ, Yang YL, Chen YM, Liu ZY. *Shiraiaceae* new family of Pleosporales (*Dothideomycetes* Ascomycota). Phytotaxa. 2013;**103**(1):51-60

[12] Hudson J, Zhou J, Chen J, Harris L, Yip L, Tpwers GHN. Hypocrellin from *Hypocrella bambuase* is phototoxic to human immunodeficiency virus. Photochemistry and Photobiology. 1994;**60**(3):253-255

[13] Index Fungorum [Internet]. 2017. Avalilable from: http://www.indexfungorum.org/Names/Names.asp [Accessed: October 20, 2017]

[14] Hino I. Illustrations fungorum bambusicolorum. Bulletin of the Miyazaki College of Agricultural Forestry. 1938;**10**(1):55-64

[15] Tanaka K, Harada Y. Bambusicolous fungi in Japan (1): Four *Phaeosphaeria* species. Mycoscience. 2004;**45**(6):377-382

[16] Dai DQ, Phookamsak R, Wijayawardene NN, Li WJ, Bhat DJ, Mortimer PE, Xu JC, Taylor JE, Hyde KD, Chukeatirote E. Bambusicolous fungi. Fungal Diversity. 2017;**82**:1-105

[17] Léveillé JH. Champignons exotiques. Annales des Sciences Naturelles ser. 1845;**3**(3):38-71

[18] Léveillé JH. Descriptions des champignons de l'herbier du Muséum de Paris. Annales des Sciences Naturelles Botanique. 1846;**5**(3):249-305

[19] Berkeley MJ, Curtis MA. Exotic fungi from the Schweinitzian Herbarium principally from Surinam. Journal of the Academy of Natural Sciences of Philadelphia. 1853;**2**(2):277-294

[20] Montagne JPFC. Cryptogamia Guyanensis seu plantarum cellularium in Guyana gallica annis 1835-1849 a clLeprieur collectarum enumeration universalis. Annales des Sciences Naturelles Botanique. 1855;**3**:91-144

[21] Hino I, Katumoto K. Illustrationes fungorum bambusicolorum II. Bulletin of the Faculty of Agriculture Yamaguti University. 1954;**5**(3):213-234

[22] Hino I, Katumoto K. Illustrationes fungorum bambusicolorum III. Bulletin of the Faculty of Agriculture Yamaguti University. 1955;**6**:29-68

[23] Hino I, Katumoto K. Illustrationes fungorum bambusicolorum IV. Bulletin of the Faculty of Agriculture Yamaguti University. 1956;**7**:267-274

[24] Hino I, Katumoto K. Illustrationes fungorum bambusicolorum V. Bulletin of the Faculty of Agriculture Yamaguti University. 1957;**8**:649-658

[25] Hino I, Katumoto K. Illustrationes fungorum bambusicolorum VI. Bulletin of the Faculty of Agriculture Yamaguti University. 1958;**9**:877-908

[26] Hino I, Katumoto K. Illustrationes fungorum bambusicolorum VII. Bulletin of the Faculty of Agriculture Yamaguti University. 1959;**10**:1175-1194

[27] Hino I, Katumoto K. Illustrationes fungorum bambusicolorum IX. Bulletin of the Faculty of Agriculture Yamaguti University. 1961;**12**:151-162

[28] Hino I, Katumoto K. Notes on bambusicolous fungi (1). Journal of Japanese Botany. 1965;**40**:81-89

[29] Hino I, Katumoto K. Notes on bambusicolous fungi (2). Journal of Japanese Botany. 1966;**41**:292-297

[30] Hino I. Icones fungorum bambusicolorum japonicorum. Japan: The Fuji Bamboo Garden Kobe; 1961

[31] Petrini O, Candoussau F, Petrini L. Bambusicolous fungi collected in southwestern France 1982-1989. Mycologia Helvetica. 1989;**3**(3):263-279

[32] Eriksson O, Yue JZ. Bambusicolous pyrenomycetes an annotated check-list. Myconet. 1998;**1**(2):25-78

[33] Zhang LQ, Wang XG. Fungus resource of bamboos in China. Journal of Bamboo Research. 1999;**18**(1):66-72

[34] Zhou DQ, Hyde KD, Vrijmoed LLP. Resources and diversity of bambusicolous fungi in China. Guizhou Science. 2000;**18**(1-2):62-70

[35] Cai L, Zhang K, McKenzie EH, Hyde KD. Freshwater fungi from bamboo and wood submerged in the Liput River in the Philippines. Fungal Diversity. 2003;**13**(1):1-12

[36] Tanaka K, Harada Y. Pleosporales in Japan (1): The genus *Lophiostoma*. Mycoscience. 2003;**44**(2):85-96

[37] Tanaka K, Harada Y. Pleosporales in Japan (3): The genus *Massarina*. Mycoscience. 2003;**44**(3):173-185

[38] Tanaka K, Harada Y. Bambusicolous fungi in Japan (4): A new combination *Astrosphaeriella aggregate*. Mycoscience. 2005;**46**(2):114-118

[39] Tanaka K, Harada Y. Bambusicolous fungi in Japan (6): *Katumotoa* a new genus of *phaeosphaeriaceous* ascomycetes. Mycoscience. 2005;**46**(5):313-318

[40] Tanaka K, Harada Y, Barr ME. Bambusicolous fungi in Japan (3): A new combination *Kalmusia scabrispora*. Mycoscience. 2005;**46**(2):110-113

[41] Hatakeyama S, Tanaka K, Harada Y. Bambusicolous fungi in Japan (7): A new coelomycetous genus *Versicolorisporium*. Mycoscience. 2008;**49**(3):211-214

[42] Sato G, Tanaka K, Hosoya T. Bambusicolous fungi in Japan (8): A new species of *Pseudolachnella* from Yakushima Island southern Japan. Mycoscience. 2008;**49**(6):392-394

[43] Morakotkarn D, Kawasaki H, Seki T. Molecular diversity of bamboo-associated fungi isolated from Japan. FEMS Microbiology Letters. 2007;**266**:10-19

[44] Tanaka K, Hirayama K, Yonezawa H, Hatakeyama S, Harada Y, Sano T, Shirouzu T, Hosoya T. Molecular taxonomy of bambusicolous fungi *Tetraplosphaeriaceae* a new pleosporalean family with *Tetraploa*-like anamorphs. Studies in Mycology. 2009;**64**(1):175-209

[45] Dai DQ, Bhat DJ, Liu JK, Chukeatirote E, Zhao RL, Hyde KD. *Bambusicola* a new genus from bamboo with asexual and sexual morphs. Cryptogamie-Mycologie. 2012;**33**(3):363-379

[46] Dai DQ, Wijayawardene NN, Bhat DJ, Chukeatirote E, Bahkali AH, Zhao RL, Xu JC, Hyde KD. *Pustulomyces* gennovaccommodated in *Diaporthaceae Diaporthales* as revealed by morphology and molecular analyses. Cryptogamie-Mycologie. 2014;**35**(1):63-72

[47] Dai DQ, Wijayawardene NN, Bhat DJ, Chukeatirote E, Zhao RL, Wang Y, Bahkali AH, Hyde KD. The phylogenetic placement of *Eriosporella bambusicola* spnovin *Capnodiales*. Cryptogamie-Mycologie. 2014;**35**(1):41-49

[48] Dai DQ, Bahkali AH, Li QR, Bhat DJ, Wijayawardene NN, Li WJ, Chukeatirote E, Zhao RL, Xu JC, Hyde KD. *Vamsapriya* (*Xylariaceae*) re-described with two new species and molecular sequence data. Cryptogamie-Mycologie. 2014;**35**(4):339-357

[49] Dai DQ, Bahkali AH, Li WJ, Bhat DJ, Zhao RL, Hyde KD. *Bambusicola loculata* sp. nov. (*Bambusicolaceae*) from bamboo. Phytotaxa. 2015;**213**(2):122-130

[50] Dai DQ, Bahkali AH, Ariyawansa HA, Li WJ, Bhat JD, Zhao RL, Mortimer PE, Xu JC, Hyde KD. *Neokalmusia didymospora* sp. nov. (*Didymosphaeriaceae*) from bamboo. Sydowia. 2016;**68**:17-35

[51] Dai DQ, Jiang HB, Tang LZ, Bhat DJ. Two new species of Arthrinium (Apiosporaceae Xylariales) associated with bamboo from Yunnan China. Mycosphere. 2016;**7**(9):1332-1345

[52] Tanaka K, Hirayama K, Yonezawa H, Sato G, Toriyabe A, Kudo H, Hashimot A, Matsumura M, Harada Y, Kurihara Y, Shirouzu T, Hosoya T. Revision of the *Massarineae* (*Pleosporales Dothideomycetes*). Studies in Mycology. 2015;**82**(2):75-136

[53] LiuJK, HydeKD, GarethEBG, AriyawansaHA, BhatDJ, BoonmeeS, MaharachchikumburaS, McKenzie EHC, Phookamsak R, Phukhamsakda RA, et al. Fungal diversityty notes 1-110: Taxonomic and phylogenetic contributions to fungal species. Fungal Diversity. 2015;**72**(1):1-197

[54] Hyde KD, Jones EBG, Liu JK, Ariyawansa H, Boehm E, Boonmee S, Braun U, Chomnunti P, Crous PW, Dai DQ, et al. Families of Dothideomycetes. Fungal Diversity. 2013;**63**(1):1-313

[55] Phookamsak R, Norphanphoun C, Tanaka K, Dai DQ, Luo ZL, Liu JK, Su HY, Bhat DJ, Bahkali AH, Mortimer PE, Xu JC, Hyde KD. Towards a natural classification of *Astrosphaeriella*-like species; introducing *Astrosphaeriellaceae* and *Pseudoastrosphaeriellaceae* famnovand *Astrosphaeriellopsis* gen. nov. Fungal Diversity. 2015;**74**(1):143-197

[56] Liu JK, Phookamsak R, Dai DQ, Tanaka K, Jones EBG, Xu JC, Chukeatirote E, Hyde KD. *Roussoellaceae* a new pleosporalean family to accommodate the genera *Neoroussoella* gen. nov. *Roussoella* and *Roussoellopsis*. Phytotaxa. 2014;**181**(1):1-33

[57] Jaklitsch W, MFournier J, Dai DQ, Hyde KD, Voglmayr H. *Valsaria* and the *Valsariales*. Fungal Diversity. 2015;**73**(1):159-202

[58] Adamčík S, Cai L, Chakraborty D, Chen XH, Cotter HVT, Dai DQ, Dai YC, Das K, Deng CY, Ghobad-Nejhad M. Fungal biodiversity profiles 1-10. Cryptogamie-Mycologie. 2015;**36**(2): 121-166

[59] Liu JK, Phookamsak R, Doilom M, Wikee S, Li YM, Ariyawansha HA, Boonmee S, Chomnunti P, Dai DQ, Bhat JD, Romero AI, Zhuang WY, Monkai J, Jones EBG, Chukeatirote E, Ko-Ko WT, Zhao YC, Wang Y, Hyde KD. Towards a natural classification of *Botryosphaeriales*. Fungal Diversity. 2012;**57**(1):149-210

[60] Ariyawansa HA, Hyde KD, Jayasiri SC, Buyck B, Chethana KWT, Dai DQ, Dai YC, Daranagama DA, Jayawardena RS, Lücking R, et al. *Fungal Diversity* notes 111-252—Taxonomic and phylogenetic contributions to fungal taxa. Fungal Diversity. 2015;**75**(1): 27-274

[61] Zhao KN, Yin GH, Zhao GZ, Cao AX. *Embryonispora* a new genus of hyphomycetes from China. Mycotaxon. 2014;**126**(1):77-81

[62] Thambugala KM, Wanasinghe DN, Phillips AJL, Camporesi E, Bulgakov TS, Phukhamsakda C, Ariyawansa HA, Goonasekara ID, Phookamsak R, Dissanayake A, Tennakoon DS, Tibpromma S, Chen YY, Liu ZY, Hyde KD. Mycosphere notes 1-50: Grass (Poaceae) inhabiting Dothideomycetes. Mycosphere. 2017;**8**(4):697-796

[63] Seifert K, Morgan-Jones G, Gams W, Kendrick B. The Genera of Hyphomycetes. Utrecht Netherlands: CBS-KNAW Fungal Biodiversity Centre; 2011. 10 p

[64] Wijayawardene NN, Hyde KD, Wanasinghe DN, Papizadeh M, Camporesi E, Bhat DJ, McKenzie EHC, Phillips AJL, Diederich P, Tanaka K, Li WJ, Tangthirasunun N, Phookamsak R, Dai DQ, Dissanayake AJ, Weerakoon G, Maharachchikumbura SSN, Goonasekara I, Hashimoto A, Matsumura M, Wang Y. Taxonomy and phylogeny of dematiaceous coelomycetes. Fungal Diversity. 2016;**77**(1):1-316

[65] Bhat DJ. Fascinating Microfungi (Hyphomycetes) of Western Ghats–India. Goa: Broadway Book Centre; 2010. pp. 1-221

[66] Kuhnert E, Surup F, Sir EB, Lambert C, Hyde KD, Hladki AI, Romero AI, Stadler M. Lenormandins A—G new azaphilones from *Hypoxylon lenormandii* and *Hypoxylon jaklitschii* spnovrecognised by chemotaxonomic data. Fungal Diversity. 2015;**71**(1):165-184

[67] Zhou CL, Wu XQ, Ji J, Ye JR. Research progress of bamboo diseases. Journal of Foresty Engineering. 2010;**24**(5):8-13

[68] Li HP. Study on Biological Characteristics and Manual Cultivation Technology of *Shiraia bambusicola* [Thesis]. Zhejiang Hangzhou: Zhejiang University; 2002

Chapter 11

Growth Characteristics of Dwarf Bamboo Distributed in the Northern Part of Japan

Masazumi Kayama and Takayoshi Koike

Additional information is available at the end of the chapter

http://dx.doi.org/10.5772/intechopen.68541

Abstract

Dwarf bamboo is a dominant forest floor species, especially in the northern part of Japan. *Sasa kurilensis, Sasa senanensis* and *Sasa nipponica* are widely distributed in this region. Growth characteristics of these three *Sasa* species are also different: leaf longevity of *S. kurilensis* is 3–5 years. In contrast, leaf longevity of *S. senanensis* and *S. nipponica* are 2 years and <1 year, respectively. We predicted that ecophysiological characteristics of the three *Sasa* species would reflect their leaf longevity; however, their characteristics were still not well analysed. We examined ecophysiological parameters of the three *Sasa* species grown under the same environment. Net photosynthetic rate at light saturation (P_{sat}) and nitrogen concentration (N) of S. nipponica showed high values after flushing. However, culms of *S. nipponica* were dropped after overwintering, and P_{sat} of the 2-year-old leaves drastically decreased. Meanwhile, P_{sat} of the current leaves of *S. kurilensis* was lower than the other two species. However, P_{sat} of 2-year-old leaves of *S. kurilensis* still maintained a relatively high value. P_{sat} of the current leaves of *S. senanensis* was higher than that of *S. kurilensis* even though N was the same. From these results, *S. senanensis* had a high photosynthetic nitrogen efficiency rate (P_{sat}/N).

Keywords: *Sasa*, photosynthetic capacity, nitrogen, chlorophyll, leaf thickness

1. Introduction

1.1. Dwarf bamboo in Japan

Dwarf bamboo is a small size bamboo species that distributes in Eastern Asia. On the classification of bamboo in Japan, dwarf bamboo is separated from the other bamboos. For dwarf bamboo, the sheath of culm remains until its death, whereas other bamboo species are removed their sheath during the process of culm growth. Six genera and 72 species of dwarf bamboo are grown in Japan (**Table 1**) [1]. Moreover, the genera of *Sasa* and *Pleioblastus* are divided into several sections by their morphological characteristics [1]. The northern limit of distribution area of dwarf bamboo is considered at the middle part of Sakhalin where the genus *Sasa* distributes [3].

The habitats of dwarf bamboo in Japan are restricted by climate, especially winter [4]. The climate in Japan is divided by the coasts of the Sea of Japan and Pacific Ocean [5], and the coast of the Sea of Japan area is considered as the snowy area. The presence of snow is an important factor to restrict the distribution of dwarf bamboo, and the main genus to distribute in the snowy area is *Sasa* [1]. Among the five sections of *Sasa* genus, *Macrochlamys* and *Sasa* can distribute in snowy area, since they are adapted to the snowy environment [1]. In this article, we introduce the ecological characteristics of the three species in genus *Sasa* that grow in the northern part of Japan. We also focus on growth characteristics of the three dominant species of genus *Sasa* with an ecophysiological method.

1.2. Ecological characteristics of three *Sasa* species in Northern Japan

In Northern Japan, the dwarf bamboo is a typical and essential component of the forest floor [6, 7]. In this region, *Sasa kurilensis*, *Sasa senanensis*, and *Sasa nipponica* distributed

Genera and section	Number of species	Typical species
Sasa		
Section *Macrochlamys*	6	*Sasa kurilensis* Makino et Shibata
Section *Lasioderma*	9	*Sasa shimidzuana* Makino
Section *Monilicladae*	4	*Sasa tsuboiana* Makino
Section *Sasa*	9	*Sasa senanensis* Rehder
Section *Crassinodi*	8	*Sasa nipponica* Makino et Shibata
Sasaella	10	*Sasaella ramosa* Makino
Sasamorpha	2	*Sasamorpha purpurascens* (Hechel) Makino
Pseudosasa	2	*Pseudosasa japonica* Makino
Pleioblastus		
Subgen. Pleioblastus	4	*Pleioblastus linearis* Nakai
Subgen. Nipponocalamus		
Section *Medakea*	6	*Pleioblastus simonii* Nakai
Section *Nezasa*	11	*Pleioblastus chino* Makino
Chimonobambusa	1	*Chimonobambusa marmoreal* (Mitford) Makino

Table 1. Genera of dwarf bamboo distributed in Japan [1, 2].

widely [1, 8]. In Hokkaido Island, which is located in the most northern part of Japan, the distribution of these three species is separated (**Figure 1**, [8]). These species are separated by the snow depth. The main distribution area of *S. nipponica* is the eastern part of Hokkaido, which faces the coast of the Pacific Ocean. Snow depth of this area is lower than other areas (below 75 cm of maximum snow depth) [4]. The distribution of *S. kurilensis* is the mountain area with heavy snow (over 150 cm of maximum snow depth). The distribution of *S. senanensis* is at the middle range of maximum snow depth between *S. nipponica* and *S. kurilensis* (75–150 cm). Also, the three *Sasa* species have different freezing tolerance. The climate in the area of the coast of Pacific Ocean the minimum temperature is lower than −10°C, and soil freezing occurs due to low snow depth [5]. The freezing tolerance for the bud of *S. nipponica* (−10 to −15°C) is higher than *S. kurilensis* and *S. senanensis* (−5 to −10°C) [9]. As a result, bud of *S. nipponica* can survive soil freezing. On the other hand, the distribution area of *S. kurilensis* and *S. senanensis* is covered with deep snow during the winter [1, 4]. The culms of *S. kurilensis* and *S. senanensis* are laid on the ground by the weight of snow cover. Snow has low thermal conductivity [10], and low air temperature is not easily conducted to the soil; as a result, soil can escape freezing [11]. Thus, *S. kurilensis* and *S. senanensis* can survive under the snow cover in winter. When the leaf and culm of *S. kurilensis* and *S. senanensis* are exposed above the snow depth, these organs cannot survive [12, 13].

The three *Sasa* species have different morphology types (**Figure 2**, [4]). The culm height of *S. kulinensis* reaches 3 m, and its longevity is estimated to be over 10 years. The upper part

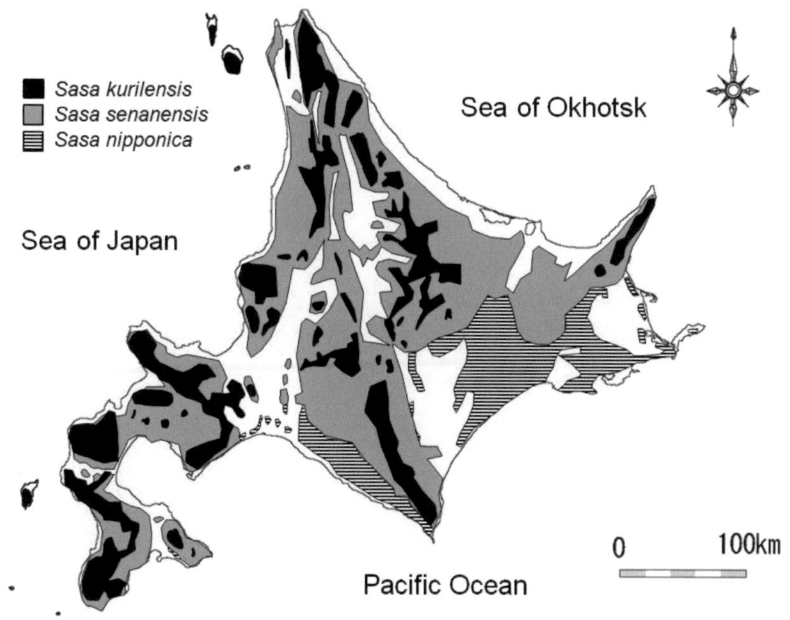

Figure 1. The distribution of the three *Sasa* species in Hokkaido Island located in Northern Japan (modification from Toyooka et al. [8]).

of the culm of *S. kulinensis* has buds on nodes and continues to bifurcate. In contrast, there are no buds at lower part of the culm of *S. kulinensis*. The leaves of *S. kulinensis* can survive for relatively longer time, and its longevity ranges from 3 to 5 years. The culm height of *S. senanensis* is about 2 m, and its longevity is 5 years. The culm of *S. senanensis* has buds on every node. The leaf longevity of *S. senanensis* is about 2 years. In contrast, the culm height of *S. nipponica* is less than 1 m, and its longevity is also about 1 year. The buds of *S. nipponica* exist at the underground of the culm. The leaf longevity of *S. nipponica* is less than 1 year.

The three *Sasa* species have well-developed rhizome systems and are dominant at the forest floor in general forests of this region [14, 15]. As a result, the light environment under the *Sasa* species is quite dark, and regeneration of other species is suppressed [16]. Moreover, the *Sasa* species has high regeneration ability after disturbances. When forests suffer from forest fires, forest cannot restore; however, dwarf bamboo is able to regenerate as ground vegetation [17, 18]. The flowering period of the *Sasa* species is estimated to be 60–100 years [4]; however, information of flowering is still limited. Based on previous information, flowering of *Sasa* species occurs synchronously and often expands over 1000 ha in area [4, 19]. After flowering, numerous seeds are produced, and all culms of the *Sasa* species dies [4, 19], as does a monocarpic plant.

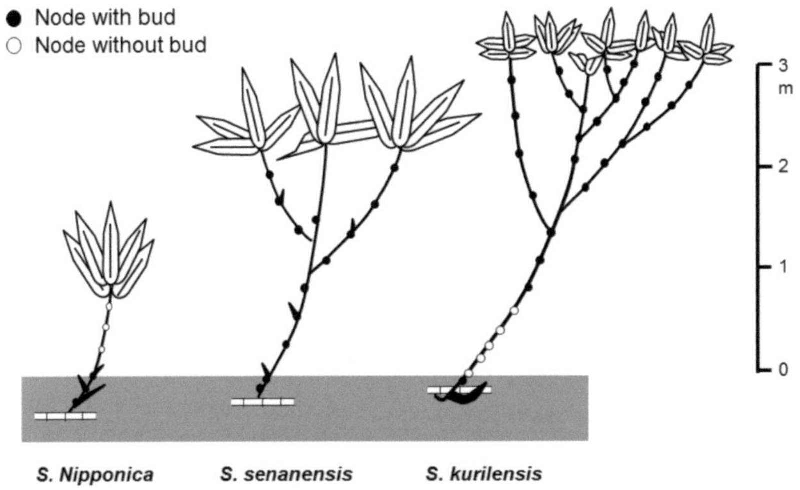

Figure 2. Morphological characteristics of the three *Sasa* species (modification from Makita [4]).

2. Ecophysiological characteristics of three *Sasa* species

2.1. Background

In the previous chapter, we summarized the ecological characteristics of the three dominant *Sasa* sp. in Northern Japan: *S. kurilensis*, *S. senanensis*, and *S. nipponica*. We showed specific traits of leaves and culm longevity of the three species. To survive and grow under different growth conditions, the *Sasa* species have adapted to each habitat through morphological and physiological adaptation. For example, leaf and stem longevities of *S. nipponica* are 1 year, and so its leaf has to obtain large amount of photosynthetic productivity during the one growing period. In contrast, leaves and culms of *S. kurilensis* can survive for a long period. Therefore, it also may be possible for *S. kurilensis* to obtain photosynthetic productivity for a long period. In general, plant growth form can be evaluated through ecophysiological characteristics [20, 21]. Photosynthetic characteristics of three *Sasa* species have been measured by previous research [22–24]. However, characteristics cannot be compared because the measurement was done under different conditions.

There are contrasting growth characteristics, namely fast and slow [25]. Fast-growing species have short-lived leaves with a high photosynthetic capacity, whereas slow-growing species have long lived leaves with a low photosynthetic capacity that can maintain its function over long periods. The differences of photosynthetic capacity between fast- and slow-growing species are related to foliar nitrogen concentration, which is usually higher in fast-growing than in slow-growing species [25]. In contrast, photosynthetic nitrogen use efficiency is an indicator for allocation of nitrogen to photosynthetic apparatus; slow-growing species shows a high value [25]. The nitrogen use characteristic is predicted to be different according to life form. For example, the photosynthetic rate and concentration of nitrogen may be high for *S. nipponica,* since it has a short leaf longevity. We predicted, in contrary, that leaves of *S. kurilensis* may have a low photosynthetic rate and low nitrogen concentration. The long longevity of *S. kurilensis* may be compensated with low photosynthetic productivity as found in several kinds of evergreen spruce [21].

The aim of this chapter is to show ecophysiological characteristics of the three *Sasa* species in relation to their different life forms, such as leaf longevity, culm height, etc. We measured the seasonal change of photosynthetic rates, concentrations of nitrogen and chlorophyll, and leaf thickness of different aged leaves of the three *Sasa* species planted in a common garden.

2.2. Materials and methods

This research was conducted in an arboretum of the Hokkaido Research Center, Forestry and Forest products Research Institute (43°00′N, 141°23′E, 141 m a.s.l.) located in Sapporo City, Hokkaido, Japan. The annual mean, maximum, and minimum temperatures at the metrological station of this centre were 7.3, 35.7, and −22.8°C, respectively, from 1975 to 2003 [26]. The range of annual precipitation was from 581 to 1490 mm year^{-1} during 1975–2003 [26]. The maximum snow depth in winter was 130 cm [26]. In this arboretum, the subterranean stem of *S. kurilensis, S. senanensis*, and *S. nipponica* was planted in 1982. The size of planting area was 5 × 10 m for each *Sasa* species. Plantations of *Sasa* species were exposed to full sunlight the whole day because there were no surrounding trees around the plantation.

We measured the photosynthetic rate at light saturation (P_{sat}, μmol m^{-2} s^{-1}) from May to October 2004. The measurements were carried out at 10:00–15:00 each month. Second leaves counted from the top of culm of each *Sasa* species were used for the measurement of P_{sat}. We selected four leaves of current and 2-year-old ones located at sunny positions. Measurements were made by using a portable gas analyzer (LI-6400, LI-COR Biosciences, Lincoln, NE, USA) under steady-state conditions (25°C, 36.0 Pa of CO_2, and 1800 μmol m^{-2} s^{-1} photosynthetic photon flux using LED), which were previously determined [27].

After measuring photosynthetic rate, we sampled the leaves and analysed the chlorophyll concentration. The fresh mass of leaves were first measured, then crushed by liquid nitrogen, and finally extracted by dimethyl sulfoxide. Measurement of chlorophyll was done by a spectrophotometer (V560, JASCO Co., Tokyo, Japan), and its concentration was calculated by an equation [28]. The remaining leaf samples were dried at 80°C, for 4 days. After drying, we measured specific leaf area (SLA = leaf area per dry mass, cm^2 g^{-1}, [29]). Leaf samples were ground to a fine powder using a sample mill (WB-1; Osaka Chemical Co., Osaka, Japan). The mass-based concentration of nitrogen (N_{mass}, mmol g^{-1}) was analysed using a NC analyser (NC-800, Sumika Chemical Analysis Service, Osaka). We also calculated the photosynthetic nitrogen use efficiency (PNUE, nmol $mmol^{-1}$ s^{-1}, [25]) as an indicator of photosynthetic apparatus allocation. PNUE was calculated by the following Eq. (1). We also calculated area-based concentration of nitrogen (N_{area}, mmol m^{-2}) from the value of SLA Eq. (2).

$$\text{PNUE} = P_{sat} / N_{area} \times 1,000 \quad (1)$$

$$N_{area} = 10,000 / \text{SLA} \times N_{mass} \quad (2)$$

The value of P_{sat}, SLA, concentrations of chlorophyll and nitrogen, and PNUE was examined using Tukey tests. The mean values were compared among *S. kurilensis*, *S. senanensis*, and *S. nipponica*.

2.3. Results

Concerning the value of P_{sat} for the current leaves, *S. nipponica* showed high values (14 μmol m^{-2} s^{-1}) from June when its leaves flushed (**Figure 3**). In July, P_{sat} of *S. nipponica* increased to 18 μmol m^{-2} s^{-1}, and its value was significantly higher than other *Sasa* species ($P < 0.01$).

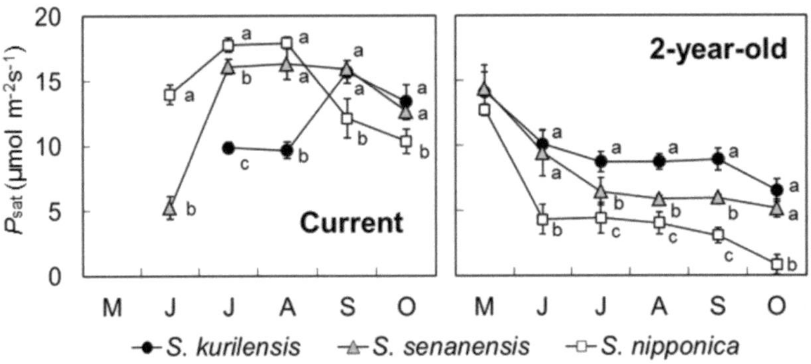

Figure 3. Seasonal change of photosynthetic rate at light saturation (P_{sat}) for current and 2-year-old leaves of the three *Sasa* species (May to October 2004, n = 4). Different letters indicate significant differences as calculated by Tukey test ($P < 0.05$).

However, P_{sat} of *S. nipponica* started to decrease from September. P_{sat} of *S. senanensis* in June was significantly lower than *S. nipponica* ($P < 0.001$); however, P_{sat} increased to 16 μmol m^{-2} s^{-1} from July to September. In contrast, flushing of leaves of *S. kurilensis* was in July, and P_{sat} was significantly lower in July and August than other *Sasa* species ($P < 0.001$). In September, P_{sat} of *S. kurilensis* increased to 15 μmol m^{-2} s^{-1}.

In 2-year-old leaves, all *Sasa* species showed high values of P_{sat} in May when all species had not yet flushed new leaves. However, P_{sat} of 2-year-old leaves was decreased from June. Especially, the culms of *S. nipponica* fell to the ground in June, and P_{sat} was drastically decreased. P_{sat} of *S. senanensis* and *S. kurilensis* was also decreased from July. However, P_{sat} of *S. kurilensis* was maintained at 8 μmol m^{-2} s^{-1} until September, and these values were significantly higher in July, August, and September than that of *S. senanensis* ($P < 0.05$).

The value of SLA was also different among the three *Sasa* species. For the current leaves, SLA in July and September showed significantly high values for *S. nipponica* than that for *S. senanensis* and *S. kurilensis* (**Figure 4**, $P < 0.05$). In contrast, SLA for current leaves of *S. kurilensis* was the lowest values from July to September. The values of SLA for current leaves decreased by time for all *Sasa* species. Compared to current and 2-year-old leaves, SLA showed low values for 2-year-old leaves for all *Sasa* species. From July, there was no significant difference in SLA of 2-year-old leaves among the three *Sasa* species.

Concentration of mass-based nitrogen (N_{mass}) in current leaves showed the highest values in June for *S. nipponica* and *S. senanensis*; however, their values decreased by time (**Figure 5**). In the case of *S. kurilensis*, the decrease of N_{mass} in current leaves was not clear. N_{mass} for current leaves was significantly higher for *S. nipponica* from June to September than those for *S. senanensis* and *S. kurilensis* ($P < 0.05$). In October, N_{mass} of *S. nipponica* showed similar value with *S. kurilensis*. As for the trend of 2-year-old leaves, all *Sasa* species decreased N_{mass} with time. N_{mass} in June of 2-year-old leaves of *S. nipponica* was significantly lower than the N_{mass} of *S. senanensis* and *S. kurilensis* ($P < 0.05$).

Compared with N_{mass}, area-based nitrogen (N_{area}) showed that its decrease by time was not obvious for current leaves. The peak of N_{area} showed in June of 2-year-old leaves for *S. kurilensis*

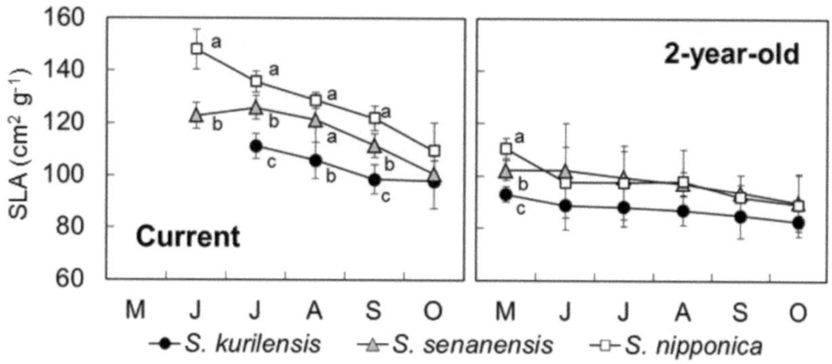

Figure 4. Seasonal change of specific leaf area (SLA) for current and 2-year-old leaves of the three *Sasa* species (May to October 2004, n = 4). Different letters indicate significant differences as calculated by Tukey test ($P < 0.05$).

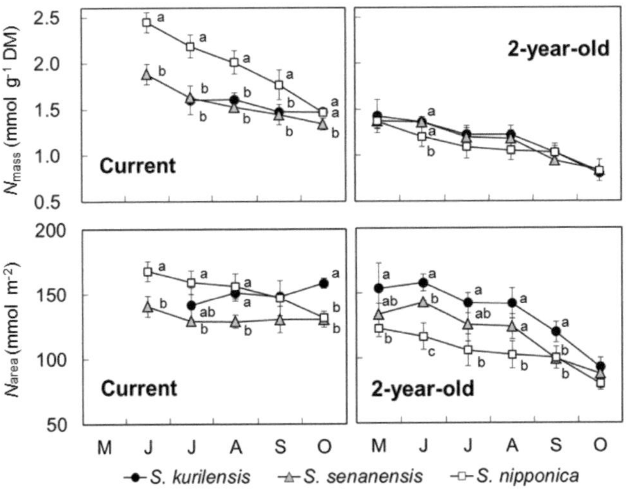

Figure 5. Seasonal change of mass-based (N_{mass}) and area-based (N_{area}) concentrations of nitrogen for current and 2-year-old leaves of the three *Sasa* species (May to October 2004, n = 4). Different letters indicate significant differences as calculated by Tukey test ($P < 0.05$).

and *S. senanensis*, whereas its peak was in June of current leaves for *S. nipponica*. N_{area} for current leaves of *S. nipponica* showed significantly higher than that of *S. senanensis* from June to August ($P < 0.01$). In contrast, N_{area} for current leaves of *S. nipponica* did not show significant difference with *S. kurilensis* from July to September. In October, N_{area} for current leaves showed significantly higher for *S. kurilensis* than those of other *Sasa* species ($P < 0.001$). Also, N_{area} for 2-year-old leaves showed significantly higher for *S. kurilensis* than that for *S. nipponica* ($P < 0.01$). N_{area} for 2-year-old of *S. senanensis* showed middle range between *S. kurilensis* and *S. nipponica*, and its trend was similar with *S. kurilensis*.

Total chlorophyll (Chl a+b) concentration showed the low value after flushing and increased in August for *S. kurilensis* and *S. nipponica* and in June for *S. senanensis* (**Figure 6**). Compared with *Sasa* species, chlorophyll concentration was significantly high value for current leaves of *S. kurilensis* in September and October (**Figure 6**, $P < 0.05$). In August, chlorophyll concentration of current leaves was significantly higher at *S. nipponica* compared to *S. kurilensis* and *S. senanensis* ($P < 0.05$).

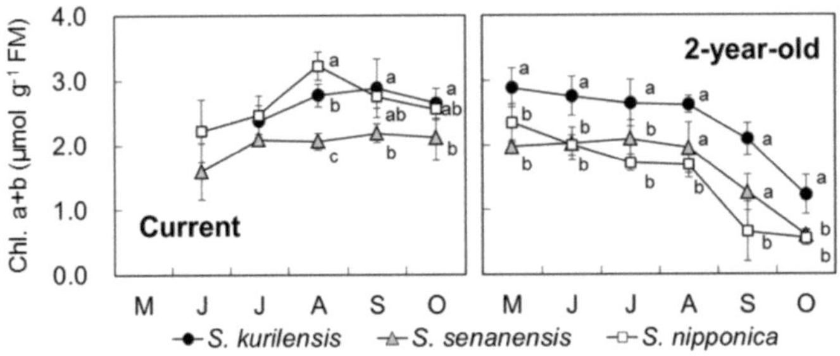

Figure 6. Seasonal change of chlorophyll (a + b) concentration for current and 2-year-old leaves of the three *Sasa* species (May to October 2004, n = 4). Different letters indicate significant differences as calculated by Tukey test ($P < 0.05$).

Chlorophyll concentration for 2-year-old leaves of *S. kurilensis* and *S. senanensis* was maintained these values compared with current leaves, whereas its value of *S. nipponica* was decreased gradually. *S. kurilensis* had a significantly higher chlorophyll concentration in all months than that of S. *nipponica* and *S. senanensis* ($P < 0.05$). Concentration of chlorophyll for 2-year-old leaves showed remarkable decrease from September.

PNUE of current leaves showed significantly high values for *S. senanensis* from July compared with the other *Sasa* species (**Figure 7**, $P < 0.05$). In contrast, PNUE of current leaves of *S. nipponica* decreased from September. PNUE of current leaves of *S. kurilensis* increased from September. PNUE of current leaves was decreased for all *Sasa* species in October.

For 2-year-old leaves, PNUE showed high in May; however, this value was decreased in June. From June to August, the value of PNUE was maintained these values for three *Sasa* species. In September, PNUE of *S. kurilensis* and *S. senanensis* was increased, whereas its value was decreased for *S. nipponica*. Compared with three *Sasa* species, PNUE for 2-year-old leaves of *S. kurilensis* was significantly higher from July to October than those for other *Sasa* species ($P < 0.05$).

2.4. Discussion

Based on the results, the ecophysiological characteristics of the three *Sasa* species were different. The leaf of *S. nipponica* showed high P_{sat} from flushing (**Figure 3**). The leaf of *S. nipponica* also had high N (**Figure 5**), and this made it possible to maintain a high P_{sat} concentration. Furthermore, the leaf of *S. nipponica* was thin with a high value of SLA (**Figure 4**). In general, thin leaves have a low value of CO_2 diffusive conductance [25]; as a result, thin leaves show high P_{sat}. So, the relatively thin leaf of *S. nipponica* has a big advantage to obtain a high photosynthetic rate through diffusion of CO_2 in its leaves. In contrast, 2-year-old culm of *S. nipponica* was fallen in June, and its leaves were laid at a low layer of the plantation. P_{sat} of 2-year-old leaves decreased (**Figure 3**), and the photosynthetic productivity of its leaf may have been small. However, we confirmed that the leaves of *S. nipponica* could survive over 1 year, even if the culm has fallen.

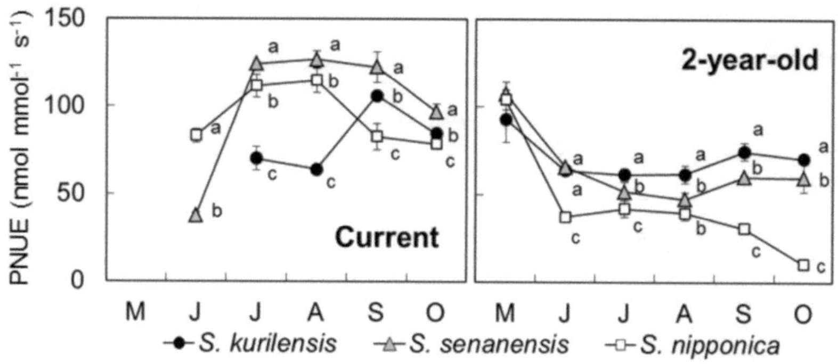

Figure 7. Seasonal change of photosynthetic nitrogen use efficiency (PNUE) for current and 2-year-old leaves of the three *Sasa* species (May to October 2004, n = 4). Different letters indicate significant differences as calculated by Tukey test ($P < 0.05$).

These characteristics that show high photosynthetic capacity and high concentration of nitrogen for younger leaf and short leaf longevity are corresponded with fast-growing species [25]. In general, fast-growing species shows that photosynthetic rate is decreased drastically by increase of leaf age [21, 30]. This trend is clear for evergreen oak compared with conifer species [30]. Moreover, there are a fast-growing species among same genus of *Picea*, and *Picea abies* and *Picea glauca* are considered as fast-growing species [21]. These species showed high photosynthetic rate for younger leaves; however, their high values were not maintained. Also, fast-growing species have a high rate of leaf turnover [31]. Woody species have a leaf turnover mechanism, and when old leaves are lost, leaf nitrogen is retranslocated to younger leaves [32]. *S. nipponica* showed continuous decrease of N_{mass} (**Figure 6**), and its trait is probably related with retranslocation of nitrogen. *S. nipponica* may be retranslocated nitrogen from old to young leaves, thus maintaining high photosynthetic capacity.

For the other *Sasa* species, the maximum value of P_{sat} for current leaves of *S. kurilensis* was lower than other species; however, its value for 2-year-old leaves was maintained for 5 months (**Figure 3**). These traits are corresponded with slow-growing species [25]. The concrete slow-growing species are *Taxus baccata, Picea mariana,* and *Picea rubens* [21, 30]. The leaf longevities of these species were over 5 years, and photosynthetic rates showed high value for 6-year-old leaves [21, 30]. Also, maximum leaf longevity of *S. kurilensis* is 5 years [4], and its ecophysiological characteristics are similar with other slow-growing species. Also, slow-growing species have a characteristic to maintain high value of PNUE for aged leaves [21, 30], and *S. kurilensis* showed high PNUE for 2-year-old leaves (**Figure 7**). This trait is related with the maintenance of photosynthetic rate for long period.

On other traits, slow-growing species has thick leaves [21, 25]. Leaves of *S. kurilensis* showed a low value of SLA (**Figure 4**), which was characterised by thick leaves. In general, species with a small SLA allocates nitrogen to the leaf cell wall and increases toughness of the cell [33]. This trait contributes to the extent of leaf longevity [34]. Thus, allocation of nitrogen in leaves for *S. kurilensis* is probably larger for cell wall than for protein of photosynthetic apparatus. As a result, *S. kurilensis* may make leaves with a long longevity but with a low photosynthetic rate.

P_{sat} of *S. senanensis* for current leaves showed high values in August and September (**Figure 3**). In contrast, current leaves of *S. senanensis* were thick (**Figure 4**), and N_{area} and N_{mass} were low compared with *S. nipponica* (**Figure 5**). Thus, ecophysiological characteristics of *S. senanensis* are not similar with *S. nipponica*. In contrast, leaves of *S. senanensis* were thin (**Figure 4**) and short longevity (about 2 years, [4]) compared with *S..kurilensis*. Thus, ecophysiological characteristics of *S. senanensis* are also not similar with *S. kurilensis*. Consequently, ecophysiological characteristics of *S. senanensis* are intermediate between fast- and slow-growing species. On the remarkable characteristics of *S. senanensis*, PNUE showed the highest value for current leaves (**Figure 7**). *S. senanensis* may allocate more nitrogen to protein of photosynthesis apparatus compared with other *Sasa* species. Similar ecophysiological characteristics were reported for *Pinus pinea* and *Picea jezoensis* var. *hondoensis* [21, 30].

In addition, the trait of chlorophyll concentration also concerns with ecophysiological characteristics. The concentration of chlorophyll showed high values for *S. kurilensis*, especially 2-year-old leaves (**Figure 6**). In general, chlorophylls have light harvesting complex proteins

(LHCP) at thylakoids in the chloroplast [35]. As an increase of chlorophyll contributes to an increase in photon absorption, chlorophyll concentration shows a positive relationship with photosynthetic rate within the same species [35]. In the case of 2-year-old leaves of *S. kurilensis*, P_{sat} showed a high value despite not having a high nitrogen concentration (**Figures 3** and **5**) and small SLA (**Figure 4**). There is a possibility that 2-year-old leaves of *S. kurilensis* allocate nitrogen to chlorophyll (Chl/N) and reinforce the absorption and transferring capacity of photon. Consequently, *S. kurilensis* may use absorbed photo efficiently for increasing photosystem by decreasing CO_2 diffusion in its leaf. Leaf longevity of *S. kurilensis* is 3–5 years [4]. Therefore, aged leaves of *S. kurilensis* are considered to be shaded by new leaves that flushed later on; therefore, mutual shading occurs. High concentration of chlorophyll in 2-year-old leaves of *S. kulinensis* may have had the advantage under shady conditions.

3. Conclusion

Sasa species regenerates at the same place with clonal development, and these traits cannot be simply classified into fast- and slow-growing species as other species. We regard the *Sasa* species as follows: *S. nipponica* is classified as a fast-growing species, whereas *S. kulinensis* are slow-growing species. Indeed, ecophysiological characteristics of *Sasa* sp. are the same as slow- and fast-growing species as found in other plant species. *S. senanensis* cannot be classified as two growing types and showed intermediate characteristics between fast- and slow-growing species.

Related to the habitat of the three *Sasa* species, edaphic habitat of *S. nipponica* is considered to be the deep humus layer and A-horizon [36]. The characteristics of a fast-growing species is to have an advantage in a fertile habitat, and the growth trait of *S. nipponica* shows a rapid turnover of leaves and culms [4], which is considered to be suitable for the habitat. We conclude that ecophysiological characteristics of *S. nipponica* are adapted to fertile habitats. The distribution area of *S. nipponica* is classified as low altitudes, facing to the coast of Pacific Ocean where the summers are relatively cloudy with high humidity and the high photosynthetic performance of *S. nipponica* is kept [8]. Moreover, although the snowy period there is short, the soil freezes with cold climate [5]. *Sasa* cannot keep evergreen leaves during winter; hence, the *Sasa* species must produce new leaves from spring after the death of leaves of previous year. Its high photosynthetic rate may be compensating short leaf longevity.

In contrast, the distribution of *S. kurilensis* is hillsides and slope of valley sides where soil depth is shallow [36]. In general, these locations restrict plant growth. The leaves and culms of *S. kurilensis* can survive for several years [4], and these traits may exist to compensate for low photosynthetic productivity. *S. kurilensis* showed high concentration of chlorophyll and PNUE for 2-year-old leaves (**Figures 6** and **7**). This characteristic is suitable for conditions where resources are limited. Thus, we conclude that ecophysiological characteristics of *S. kurilensis* reflect the adaptability to infertile habitats. *S. kurilensis* distributes at high mountain areas in Hokkaido Island (**Figure 1**). The area of *S. kurilensis* probably corresponds with deep snow and harsh environmental conditions.

The habitat of *S. senanensis* is similar to the soil condition of *S. nipponica* [36]. Leaf longevity of *S. senanensis* is about 2 years [4], and this characteristic is probably suitable for relatively good environmental conditions, such as high soil fertility. Compared with *S. nipponica*, N_{mass} and N_{area} in current leaves were lower for *S. senanensis* (**Figure 5**). Thus, the nutrient requirement of *S. senanensis* is also lower than that of *S. nipponica*, and *S. senanensis* can adapt to infertile habitats or resources limited conditions. Moreover, the longevity of culm of *S. senanensis* is about 5 years, which is different from its leaves. Its culm has buds at every node (**Figure 2**), and the leaves can flush during the latter period of the culm life-span. Based on these results, we conclude that the growth characteristics of *S. senanensis* may be high flexibility, and it is also able to adapt to different nutrient and environmental conditions. In fact, the distribution area for *S. senanensis* in Hokkaido Island is the largest (**Figure 1**). The flexibility of *S. senanensis* may be enabling this species to grow in a broad distribution range.

Acknowledgements

We thank Dr. M. Kitao, Dr. H. Tobita, Dr. H. Utsugi (FFPRI) and Prof. Y. Maruyama (currently Nihon University) for their invaluable comments on this study. Thanks are also given to researchers at the Hokkaido Research Center, FFPRI for their encouragement. We are grateful to the technical staff of Hokkaido Research Center, FFPRI for maintenance of the arboretum. We thank Ms. Saki Fujita (Hokkaido University) for English improvement and Ms. M. Ohbuchi for assistance in the experiments.

Author details

Masazumi Kayama[1,2*] and Takayoshi Koike[3]

*Address all correspondence to: kayama@affrc.go.jp

1 Hokkaido Research Center, Forestry and Forest Products Research Institute, Sapporo, Japan

2 Japan International Research Center for Agricultural Sciences, Tsukuba, Japan

3 Silviculture and Forest Ecological Studies, Hokkaido University, Sapporo, Japan

References

[1] Suzuki S. Illustrations of Japanese Bambusaceae. Funabashi: Jukai Shorin; 1996. p. 271. (in Japanese)

[2] Uchimura E. Bamboo Guide Book. Tokyo: Soshinsha; 2014. p. 269. (in Japanese)

[3] Ohrnberger D. The Bamboos of the World. Amsterdam: Elsevier Science; 1999. p. 596

[4] Makita A. The characteristics of life history for dwarf bamboo. In: Koike T editor. Tree Physiological Ecology. Tokyo: Asakura Publishing Co.; 2004. pp. 199-214. (in Japanese)

[5] Japan Meteorological Agency. General Information on Climate of Japan [Internet]. 2017. Available from: http://www.data.jma.go.jp/gmd/cpd/longfcst/en/tourist.html

[6] Ishibashi S. The relationship between natural regeneration and land/forest description in natural cool-temperature and boreal forests. Journal of the Japanese Forest Society. 1998;**80**:74-79. (in Japanese and English summary)

[7] Noguchi M, Yoshida T. Factor influencing the distribution of two co-occurring dwarf bamboo species (*Sasa lurilensis* and *S. senanensis*) in a conifer-broadleaved mixed stand in northern Hokkaido. Ecological Research. 2005;**20**:25-30

[8] Toyooka H, Sato A, Ishizuka M. Distribution Map of the *Sasa* Group in Hokkaido. Sapporo: Hokkaido Branch, Forestry and Forest Products Research Institute; 1983. p. 36. (in Japanese)

[9] Sakai A. Plant Cold Hardiness and Adaptation to Cold Climate. Tokyo: Center for Scientific Publisher; 1982. p. 469. (in Japanese)

[10] Sturm M, Perovich DK, Holmgren J. Thermal conductivity and heat transfer through the snow on the ice of the Beaufort Sea. Journal of Geophysical Research. 2002;**107**(C21):8043. DOI: 10.1029/2000JC000409

[11] Kayama M. Study on the adaptation capacity of spruce species grown on serpentine soil and its application for forest rehabilitation. Research Bulletin of the Hokkaido University Forests. 2006;**60**:33-78. (in Japanese and English summary)

[12] Sasa K, Satoh F, Komiya K, Fujiwara K. Possibility of *Sasa* grassland structure as an environmental indicator in mid winter season at cold and strong wind region. Translocation of the Meeting in Hokkaido Branch of the Japanese Forest Society. 1994;**42**:50-52. (in Japanese)

[13] Ishikawa M, Oda A, Fukami R, Kuriyama A. Factors contributing to deep supercooling capability and cold survival in dwarf bamboo (*Sasa senanensis*) leaf blades. Frontiers in Plant Science. 2015;**5**:791

[14] Yajima T, Watanabe N, Shibuya M. Change in biomass of above- and under-ground parts in *Sasa kurilensis* and *Sasa senanensis* stands with culm height. Journal of the Japanese Forest Society. 1997;**79**:234-238. (in Japanese and English summary)

[15] Takeda K, Ito T. Effect of the rhizomes of *Sasa nipponica* on the slope reinforcement for preventing surface failure II. Field survey of the structure of *Sasa nipponica* population. Journal of the Japanese Society of Revegetation Technology. 2003;**28**:431-437. (in Japanese and English summary)

[16] Noguchi M, Yoshida T. Tree regeneration in partially cut conifer-hardwood mixed forests in northern Japan: Roles of establishment substrate and dwarf bamboo. Forest Ecology and Management. 2004;**190**:335-344

[17] Takaoka S, Sasa K. Landform effects on fire behavior and post-fire regeneration in the mixed forests of northern Japan. Ecological Research. 1996;**11**:339-349

[18] Kayama M, Nomura M, Sugishita Y, Satoh F, Sasa K, Koike T. Growth and survival of Sasa (*Sasa senanensis*) community grown under cold and strong windy environment in northern Japan. Bamboo Journal. 2000;**17**:20-26

[19] Makita A. Survivorship of a monocarpic bamboo grass, *Sasa kurilensis*, during the early regeneration process after mass flowering. Ecological Research. 1992;**7**:245-254

[20] Mediavilla S, Escudero A. Photosynthetic capacity, integrated over the lifetime of a leaf, is predicted to be independent of leaf longevity in some tree species. New Phytologist. 2003;**159**:203-211

[21] Kayama M, Kitaoka S, Wang W, Choi DS, Koike T. Needle longevity, photosynthetic rate and nitrogen concentration of eight spruce taxa planted in northern Japan. Tree Physiology. 2007;**27**:1585-1593

[22] Agata W, Hakoyama S, Kuwamitsu Y. Influence of light intensity, temperature and humidity on photosynthesis and transpiration of *Sasa nipponica* and *Arundinaria pygmaea*. Botanical Magazine Tokyo. 1985;**98**:125-135

[23] Lei TT, Koike T. Functional leaf phenotypes for shaded and open environments of a dominant dwarf bamboo (*Sasa senanensis*) in northern Japan. International Journal of Plant Science. 1998;**159**:812-820

[24] Ohshima T. Ecological studies of Sasa communities III. Photosynthesis and respiration of *Sasa kurilensis*. Botanical Magazine Tokyo. 1961;**74**:349-356

[25] Lambers H, Chapin FS III, Pons TL. Plant Physiological Ecology. New York: Springer-Verlag; 1998. p. 540

[26] Hokkaido Research Center. Forestry and Forest Products Research Institute. Annual Report 2004. Sapporo: Hokkaido Branch, Forestry and Forest Products Research Institute; 2005. p. 116. (in Japanese)

[27] Kayama M, Yamanaka T. Growth characteristics of ectomycorrhizal seedlings of *Quercus glauca, Quercus salicina*, and *Castanopsis cuspidata* planted on acidic soil. Trees. 2014;**28**:569-583

[28] Barnes JD, Balaguer L, Manrique E, Elvira S, Davison A. A reappraisal of the use of DMSO for the extraction and determination of chlorophylls a and b in lichens and higher plants. Environmental and Experimental Botany. 1992;**32**:85-100

[29] Reich PB, Walters MB, Ellsworth DS. Leaf life-span in relation to leaf, plant, and stand characteristics among diverse ecosystems. Ecological Monograph. 1992;**62**:365-392

[30] Escudero A, Mediavilla S. Decline in photosynthetic nitrogen use efficiency with leaf age and nitrogen resorption as determinants of leaf life span. Journal of Ecology 2003;**91**:880-889

[31] Aerts R, Chapin FS III. The mineral nutrition of wild plants revisited: A re-evaluation of processes and patterns. Advance of Ecological Research. 2000;**30**:1-67

[32] Killingbeck KT. Nutrient resorption. In: Noodén LD, editor. Plant Cell Death Processes. San Diego: Elsevier-Academic Press; 2004. pp. 215-226

[33] Takashima T, Hikosaka K, Hirose T. Photosynthesis or persistence: Nitrogen allocation in leaves of evergreen and deciduous *Quercus* species. Plant, Cell and Environment. 2004;**27**:1047-1054

[34] Onoda Y, Hikosaka K, Hirose T. Allocation of nitrogen to cell walls decreases photosynthetic nitrogen-use efficiency. Functional Ecology. 2004;**18**:419-425

[35] Lawlor DW. Photosynthesis. 3rd ed. Oxford: BIOS Scientific Publishers Ltd.; 2001. p. 386

[36] Sakai A. The Distributions of Plants and Adaptation to Environments, from Tropical Region to Polar or Desert Region. Tokyo: Asakura Publishing Co.; 1995. p. 164. (in Japanese)